R. de Boer

The Engineer and the Scandal

*One must
have hypotheses
and theories
in order
to organize
one's knowledge:
otherwise
everything remains
mere debris*

Georg Christoph
Lichtenberg
(1742-1799)
Physicist and witty satirist
in Göttingen

Reint de Boer

The Engineer and the Scandal

A Piece of Science History

With 103 Figures

Springer

Professor Dr.-Ing. Reint de Boer
o. Professor em. für Mechanik
an der Universität Duisburg-Essen, Campus Essen
Institut für Mechanik
Fachbereich Bauwissenschaften
D-45117 Essen

http://www.uni-essen.de/mechanik/Mitarb/deBoer.htm
reint.deboer@uni-essen.de

ISBN 978-3-642-06206-3 e-ISBN 978-3-540-27303-5

Springer is a part of Springer Science+Business Media

springeronline.com

Printed in The Netherlands

Final processing by PTP-Berlin Protago-TEX-Production GmbH, Germany
Cover-Design: Erich Kirchner, Heidelberg
Printed on acid-free paper 62/3141/Yu - 5 4 3 2 1 0

Dedicated to
my wife Vera

Preface

When, in the early 1980s, investigations into the thermomechanical theory of heterogeneously composed bodies (theory of porous media) began on the part of the author and his co-workers at the University of Essen, it was soon recognized that it would be helpful for the creation of a consistent theory to follow the historical development of the theory of porous media, i.e, to collect all hitherto known results gained in the past. In this connection the author recalled some excellent papers on porous media theory published by Professor Heinrich (with Dozent Desoyer) from the Technische Hochschule in Vienna in the middle of the 1950s (Heinrich and Desoyer, 1955,1956, 1961)[1] and the beginning of the 1960s, which had been shown to the author during his sabbatical leave at the Technische Hochschule of Vienna at the beginning of the 1970s. (In addition, on this occasion the author was introduced to Professor Heinrich himself, who was at that time seventy years old). These papers have, unfortunately, not reached the international scientific community because they were written in German. Nevertheless, the papers had a high scientific standard and it was self-evident for the author to look for preceding papers of Professor Heinrich on porous media theory. However, the author's efforts remained unsuccessful for a considerable time. A lucky break brought the author, finally, on the right track. He found a small hint to a publication by Professor Heinrich (1938) in the book *Theoretical Soil Mechanics* by Professor von Terzaghi (1943), known as the father of modern soil mechanics.

In the publication, in a little known journal, Professor Heinrich thoroughly worked out the scientific fundamentals of the theory of the consolidation of clay layers. In a footnote he remarked that he intended to use the

[1] The bibliography, basis of this book, is listed at the end. Books and papers in journals, used in the description of the scandal and in the biographies are cited with the author's name and the year of publication. All other supporting documents, such a letters, notes, reports etc. are only listed at the end of the References. This is also valid for the evidence of the figures.

concept of a certain Professor Fillunger and referred to a pamphlet by Professor Fillunger (1936) *Erdbaumechanik?* (Earthwork mechanics?) edited by himself. The brochure *Erdbaumechanik?* opened the gate for unveiling one of the biggest and most dramatic scandals in science history, namely the bitter dispute between the excellent professors von Terzaghi and Fillunger from the Technische Hochschule of Vienna around the middle of the 1930s and its sad end. Furthermore, the intensive investigations have also revealed the story of the glittering life of von Terzaghi and a scientific masterpiece in the development of the theory of porous media by Fillunger, who is recognized as the founder of this theory today.

During the late 1980s and the 1990s the author investigated this piece of science history in several places in the world, namely in Austria, Norway and the United States of America. The investigations disclosed an amazing story which had been nearly completely forgotten in the world, even in Vienna, although all the newspapers in Vienna made reports about on the heavy dispute and its tragic end with large headlines in 1937.

The inquiries lasted a long time. Many files on the strange affair in the archives in Vienna and Oslo, a large number of notes and letters as well as related publications including the extended handwritten (with Latin and Gothic German letters) diaries of von Terzaghi had to be studied. Moreover, a great deal of interviews and correspondence had to be carried out. This could only be done with the financial support of several foundations, with the support of the archivists in Vienna and Oslo, collegues and my co-workers (extending acknowledgements are listed at the end of the book).

Mrs. P. Lindner-Roullé and Mrs. G. Bujna, responsible for the word processing, have brought the manuscript to its present form and Mrs. V. Jorisch has taken care of the large number of the historical photographs. I would like to express my deep gratitude to them. This is also valid for the considerable work of Mr. J. R. Campbell, who corrected my English and brought it into the right form.

I record here also my heartfelt thanks to Springer-Verlag for the careful publishing and the pleasant cooperation.

Essen, July, 2004 Reint de Boer

Contents

Prologue

This book gives one an indepth study into an important part of the development of the Theory of Porous Media as well as the amazing story of the glittering life of Professor Karl von Terzaghi and provides an outline of the bitter dispute between him and Professor Paul Fillunger, both working at the Technische Hochschule in Vienna in the fields of Soil Mechanics and Technical Mechanics, in the early to mid 1930s. The ugly confrontation with its tragic end was a scandal in many respects.

Karl von Terzaghi, considered in civil engineering circles as the father of modern soil mechanics, is still, nearly forty years after his death, a legend for many geotechnical engineers. Also in some geological circles he is known due to some contributions on problems in that field. This appreciation is surely not exclusively founded in his consulting activities, which he so successfully exercised all over the world. There have been other excellent engineers in the field of soil and structural mechanics who are widely unknown today. Rather it was his scientific achievements which made him famous. His disciple R. B. Peck stated in a speech delivered at the ASFE (The Association of Engineering Firms Practicing in the Geosciences) annual meeting, Boston, April 12, 1988: "Several milestones, perhaps three, mark the transition to what we now think of as modern soil mechanics in the United States.

The first of these was Terzaghi's establishment of the new sciences: the mathematical theory of consolidation and the accompanying recognition of effective stress, the deformation conditions controlling earth pressure, and the determination of numerical value for the pertinent physical properties of earth materials." The two other milestones were, in the opinion of Peck, Arthur Casagrande's (main disciple of von Terzaghi) installation of an effective teaching program at Harvard, and the International Conference on Soil Mechanics and Foundation Engineering held at Harvard in June, 1936.

Indeed, in particular, the description of the consolidation problem and the devolvement of the concept of effective stress have been closely associated with von Terzaghi's name up to the present day.

Karl von Terzaghi's biography is easy to follow. He left an extensive record of his life in diaries, manuscripts, books, pamphlets, statements, notes etc. In particular, his diaries contain a lot of facts about his life, individuals, who accompanied him, and his surroundings. However, von Terzaghi was a vain person and belonged to that group of people who work their whole lifetime on their own memorial. In his diaries he sometimes described important events in his life not on the day on which they occurred, but a long time later, and he glossed over many facts. Thus, one has to be careful in adopting his view on facts and his description of certain occurrences uncritically. There are several inconsistencies in his diaries, memoranda, notes, etc. In many cases his main statements have been cross-checked by the author in order to find the truth, and not only von Terzaghi's standpoint. This procedure was not always successful. However, the reader will surely be amused by many stories told by von Terzaghi, which sometimes remind one of colorful fairy tales.

Nevertheless, the biography will reveal a brilliant, fascinating man, a highly-gifted, independent engineer, an imaginative, enthusiastic researcher, a restless, fearless adventurer, a charming, humorless cosmopolitan – and sometimes a bragging showman with immense craving for recognition. Moreover, he was a very egocentric person, who mostly only accepted his own opinion (Thus, it is not surprising that he worked with only few people and not in a team). In other words, he obviously suffered from a special kind of "secondary narcism" – well known in psychoanalysis; we will find numerous examples for this narcissistic disorder in his biography.

The opposite of von Terzaghi's dazzling personality was Paul Fillunger. It begins already with the fact that he left behind only little in written form. Thus, only a short biography can be presented, which shows a man who lived in a very small world of his own. He was, in a certain sense, arrogant, stern, and sometimes a know-it-all. However, Fillunger was an excellent scientist who investigated new scientific fields.

It is well-known in science history that new scientific findings are, in general, not accepted without struggle. Ideas and personalities clash, mostly not in a rational manner, but rather in an emotional and polemic one. Examples are the great feud between Newton and Leibniz on the invention of differential calculus, Maupertuis and König on the principle of least action and Bolzmann and Zermelo on problems of thermodynamics. The bitter dispute between von Terzaghi and Fillunger arose at the beginning and in the middle of the 1930s over certain physical problems in porous bodies. The Fillunger-von-Terzaghi occurrence, with its sad end, presents a microcosm of the unrest and upheaval in Middle Europe during this period. The whole affair and its tragic results were a big scandal at the Technische Hochschule of Vienna and had far-reaching consequences in engineering science. There are a lot of documents covering the whole strange affair from the very beginning until the end and through the aftermath. However, there are still some open points which have not been able to be clarified up to now. However, the research so far has given a clear view of the whole scandal, which seems to be unique in European scientific history.

During his studies at the Technische Hochschule in Graz, Karl von Terzaghi was an active member of the duelling fraternity *Vandalia* – a particularly wild society according to its name. Karl's sometimes strange behavior in the fraternity or in meetings with his fraternity friends may cause offence for some people who were closely related to him or are admirers of his outstanding achievements. For those people a brief overview of the significance and the development of fraternities in the German-speaking countries may be helpful in order to show that Karl's sometimes wild and improper manners as well as his exuberance in his youth were not a personal weakness of character, but rather corresponds to the tradition of the fraternities.

There are only few human societies which have existed continuously for nearly 200 years and have seemingly survived all economical and social fluctuations unchanged (see Krause, 1979). The fraternities have been persecuted and banned – and high-ranking people belonged and belong to them as members. Although they are considered as old-fashioned they still exist today. They use their own vocabulary and have strange habits, show enthusiasm for ideals already declared dead and attract seemingly quite reasonable men of different ages to their ranks.

There is not a single unique fraternity. There are educational and academic fraternities, duelling and non-duelling, those that wear colors and those that don't, confessional and confessionally unbound whose various tendencies are reflected in a manifold of terms, like Burschenschaft (students' associations), Corps, Landsmannschaft (organization of German expellees), Sängerschaft (choral society), Turnerschaft (gymnasts) and Verbindung (fraternity).

A fraternity consists of the acting students and the so-called Alten Herren (alumni). Both groups have their own as well as common meetings.

At the end of the 18th Century student organizations of German expellees and orders had been formed and had reached great importance. A decisive change occurred when, on this basis, corps and student associations developed. The modern fraternities were created through both these types of incorporation.

The spirit of a fraternity is surely determined by three things: The voluntary binding of oneself to given principles, the inner democracy, characterized by the word *Convent*, and the custom described by the expression *Comment*. Four principles are common in all fraternities, namely friendship for life, fulfillment of the duties of study, the demand for an honest way of acting – and the encouragement of young men to sow their wild oats. Thus, in those days drinking, rioting and duelling were essential parts of fraternity life and "considered a legitimate and laudable manifestation of youthful exuberance, but it claimed many victims and its effects on the individuals engaged in these manifestation was unpredictable."[2] Some of them, taking full advantage of their academic liberties, "perished in the gutter, victims of alcoholism or venereal diseases." However, most of the students "turned into painfully respectable citizens of the Main Street type" or became brilliant men in their profession.

This part of fraternity life was influenced and often supported by the alumni who paid for beer and brothel visits. However, they supported the students also in very positive respects, like spending money for the fraternity and giving aid for the young people to get good positions in the professional world.

Social life has always been the center in the life of students and the members of the fraternities – of course, connected with drinking a lot of alcohol, mostly in the form of beer, which sometimes ended in drinking bouts, whereby some students were so drunk that they completely passed out. Often the drinking bouts were connected with drinking games, like the *beer-state*, the *Doctor cerevisiae et vini-* and the *pope-game* which is also mentioned in Goethe's Faust. Most famous was the so-called *Ledersprung* (leather jump) in Leoben, Austria (see Section 1.2).

It was quite natural for students to pass rioting and noisy through the streets after drinking bouts and sometimes sharp laws had to be introduced against disorderly conduct and disorder.

However, the students – at all times – liked also to sing in a style like *Carmina burana* – and they had their own songs, e.g. *Gaudeamus igitur* and the *Krambambuli* songs.

To the social meetings of the students belonged the *Kneipe*[3] which they had to attend at least once a week. They spent the evening with discussions, playing some games – and drinking.

At the turn of the century, 1900, duelling was very popular in the fraternity. Karl von Terzaghi performed the duels with great enthusiasm, as we will see in Section 1.2.

[2] Text in quotation marks consists of statements by Karl von Terzaghi, if no other reference is made. Remarks in parantheses accompanied by "the author", means the opinion of the author of this book.

[3] Kneipe means bar, but may also refer to student's drinking event

1. Karl von Terzaghi's Childhood, Youth and his Road to Practice and Theory

1.1 Childhood

Karl von Terzaghi was a descendant of a long line of Austrian professional men and army officers who lived in Lodi, Lombardy and Prague, Bohemia. His family tree can be traced back to 1762 when his great-grandfather Peter Terzaghi was born. He later became a lawyer and married Maria Leonardi. They named their son, born in 1806, Peter Anton (Pietro Antonio) (grandfather of Karl) who entered the Austrian army where he advanced to the rank of Oberleutnant later. In 1832 Peter Anton married Johanna Rischan from Prague. Peter Anton was awarded by Kaiser Franz Josef I with the title *Edler von Pontenuovo* in 1854, for "having attained the rank of Major, with more than 31 years of continuous military service and decorations for valor" (Goodman, 1999). Karl von Terzaghi's father, Antonius de Padua Petrus Aloisius Franciscus Johannes Nepomucy Edler von Pontenuovo (1839 – 1890), was stationed as a commander of an infantry battalion in Prague, and became later a Colonel (Oberstleutnant). In 1882, he married Amalie Phillipine Eberle (1853 – 1942) in Graz, the daughter of Karl Andreas Eberle (1823 – 1916) and Amalia Gettinger (1826 – 1904). Karl Andreas Eberle served in the Austrian Tobacco-Administration after he had finished his studies in mechanical engineering in 1846. In the 1870s he organized, as a general manager of an international bank consortium under extreme difficulties, the production of tobacco in Romania. He earned a lot of money and became a wealthy man.

On October 2, 1883, Karl Anton Terzaghi, Edler von Pontenuovo, was born in Prague (in the following paragraphs we refer to him simply as Karl von Terzaghi[4]).

He was born within the great Austro-Hungarian Empire, the Danube monarchy, at that time relatively quiet with only few conflicts so that he was able to spend his childhood and youth in an intact world. This is also true for his family life. His wealthy family presented him the right surroundings for his carefree development.

[4] Karl von Terzaghi used several versions of his name in the literature: Karl von Terzaghi, Karl v. Terzaghi, Karl Terzaghi, and Charles v. Terzaghi.

The Austro-Hungarian monarchy at the time of Terzaghi's birth and youth.

Karl's birthplace in Prague

Grandfather Karl Eberle's house in Graz.

Nearly one and a half years later, on February 21, 1885, Amalie gave birth to Karl's sister, Gabriele Anna Johanna Nep. Terzaghi Edle von Pontenuovo, known later as Ella.

The von Terzaghis lived in an "old house on Thungasse, from which, over the tiled rooftops, one glimpses a high wooded hillside with only a church and a tower to mar the natural landscape rising above the city" (Goodman, 1999). Von Terzaghi's home town, Prague, is also called the *Golden Town* and was at that time the capital of the Austrian Province of Bohemia. It belongs to the loveliest cities in Europe and is known as the town of Gothic and Baroque. The city is divided by the legendary river Moldau. To the left of the Moldau the Hradschin, now seat of the Czech president, the castle Belvedere, and the municipal district Kleinsseite where its old narrow lanes and baroque palaces, are located. The old Karlsbrücke, beautified with many statues, leads to the right Moldau bank with its old town, university, and Jewish quarter. At that time the cultural and social life in Prague was influenced by German culture to a great extent and the Deutsche Technische Hochschule had educated many excellent engineers in Bohemia and also in the German-speaking countries.

Little is known about the childhood of Karl von Terzaghi in Prague. Only a few anecdotes are reported. In a letter to the author the late Ruth von Terzaghi (Karl von Terzaghi's second wife) pointed out: "I do not recall that Karl ever expressed any recollections of his parents from his childhood, although he did recall with apparent pleasure an orderly of his father ... The orderly (I suppose a sergeant or the equivalent) apparently enjoyed little Karl's company and made for him a toy barn complete with animals. To get a better view of the interior, Karl stuck his head into the barn and was unable to withdraw it. The orderly of course came promptly to the rescue." The siblings (Karl and Ella) stayed only for a few years in Prague. Their father retired in 1887 and the family moved to Graz, in the Austrian state of Styria, next to Meran the preferred place of residence for retired officers. Grandfather Eberle had bought a comfortable three-story house in the Rechbauerstr. 10, which was built in 1876. The von Terzaghis lived on the first floor. The large block was located directly across from the Technische Hochschule and only a short distance from the opera house and city park.

Unfortunately, Karl's father was not able to enjoy his retirement for a long time as he died shortly thereafter in 1890. Karl and Ella von Terzaghi's maternal grandfather, Karl Eberle, seems to have been happy to act in loco parenties. He exercised great influence on Karl's development. As he himself later remarked: "His strong, distinct personality and his honorable, masculine character influenced my development" and Ruth von Terzaghi continued in her letter: "Karl had many fond memories of him. One of these centered on the freshly roasted chestnuts which Grandfather regularly bought from street vendors and brought home to his grandchildren. Later, he seems to have played, or at least tried to play, a considerable part in the choice of careers."

Karl's father:
Antonio von Terzaghi

Karl's mother:
Amalie Phillipine von Terzaghi

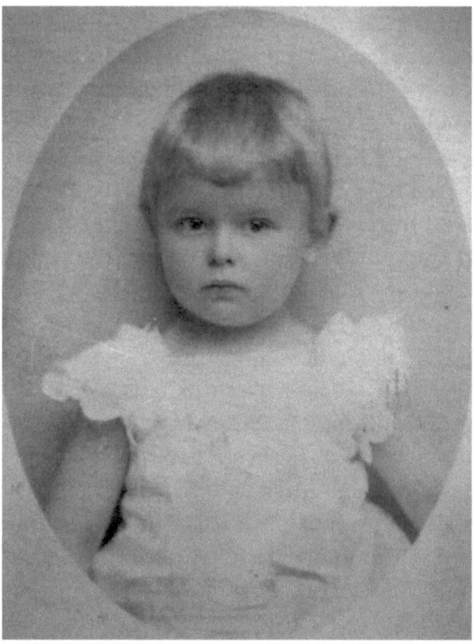

Karl as a toddler, around 1886

Ruth von Terzaghi further wrote about Karl: "As a child, he might perhaps have been described as either *mischievous* or *troublesome*. (In the US it is almost an article of faith that gifted children are troublesome). Karl recalled one occasion when he and his younger sister Ella (his only sibling) were walking along a riverbank with a nurse; noting the nurse's inattention, he hid Ella behind a tree, rolled a large stone into the water, and led the nurse to believe that Ella had fallen into the water."

It was quite natural that Karl would become a career officer following the tradition of his ancestors. In 1893 he became a pupil in the military Unterrealschule in nearby Güns (Köszeg, Hungary) with the intention of applying later for admission into the Kriegsmarine (navy). The teaching positions were exclusively filled up by officers who were able to rouse vivid interest for sciences. The spartan living and the education to bear physical hardship laid an excellent basis for the practice of Karl's later profession as a journeyman in civil engineering.

However, already in the second year at the military Unterrealschule he had a hunch that he would not adjust to the military profession. But in the time to follow he concentrated his whole interest on geography, in particular on the polar regions. He shared his interests with his lively friend Hans Kalbacher, a very talented boy who became Karl von Terzaghi's life-time friend. They ordered books, catalogues, drew maps and worked on different technical problems. Moreover Karl became interested in astronomy. Despite a lack of mathematics and mechanics he worked half a year on the approximate calculation of the sunrise and sunset.

In his 15th year of age he entered the Military Oberrealschule in Mährisch-Weisskirchen (Hranice, Czech Republic). This year was of great influence for his further development. Geography as a mere description of countries was no longer of interest to him. This field was replaced with mathematics – algebra and geometry. In February 1898 he got sick for six months. During this time he was deeply involved in geometric problems and also philosophical questions gradually arose like the questions about the purpose of human existence and the birth of the universe. The unanswered questions bothered him and he turned to religious studies. The new interest for science and the antipathy towards the boring "soldier game" led him to resolve to leave the military school. However, the relatives in Graz did not agree, at first. Finally, after struggling he succeeded in being granted his wish.

During the summer of 1898 he prepared himself for the entrance examination to the Landes-Oberrealschule in Graz. He was not able to continue all his other studies with the exception of his reading in religion. Gradually, he came to doubt many statements in the Bible and he did not know what he should believe. Finally, he regarded religion as a superfluous tool which he abandoned out of his life.

Karl entered the 6th grade of the Landes-Oberrealschule. In a short time his interest in describing natural sciences was becoming his favorite field. He

Karl with mother, Ella and grandparents about 1897

Karl at age 14

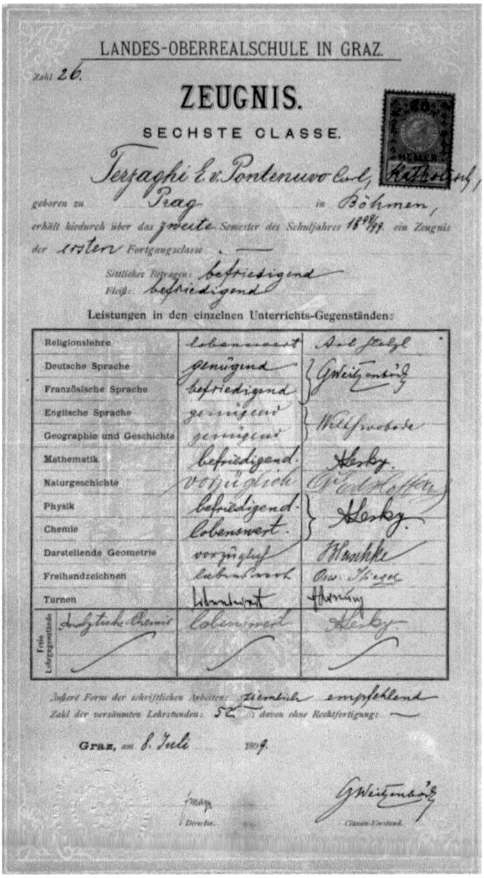

LANDES-OBERREALSCHULE IN GRAZ.

Zahl 26.

ZEUGNIS.

SECHSTE CLASSE.

Terzaghi E. v. Pontenuovo Carl, Katholisch,

geboren zu Prag in Böhmen,

erhält hiedurch über das zweite Semester des Schuljahres 1898/99 ein Zeugnis der ersten Fortgangsclasse.

Sittliches Betragen: befriedigend

Fleiß: befriedigend

Leistungen in den einzelnen Unterrichts-Gegenständen:

Religionslehre	lobenswert	[illegible]
Deutsche Sprache	genügend	[illegible]
Französische Sprache	befriedigend	
Englische Sprache	genügend	[illegible]
Geographie und Geschichte	genügend	
Mathematik	befriedigend	Herky
Naturgeschichte	vorzüglich	[illegible]
Physik	befriedigend	Herky
Chemie	lobenswert	
Darstellende Geometrie	vorzüglich	[illegible]
Freihandzeichnen	lobenswert	[illegible]
Turnen	lobenswert	[illegible]
Freie Lehrgegenstände: Analytische Chemie	lobenswert	Herky

Äußere Form der schriftlichen Arbeiten: ziemlich empfehlend

Zahl der versäumten Lehrstunden: 52, davon ohne Rechtfertigung: —

Graz, am 8. Juli 1899.

Director. Classen-Vorstand.

Karl's school report from the 6th class
at the Landes-Oberrealschule in Graz.

studied intensively the books recommended by his teacher, in particular Darwin's *The origin of the species*. Ruth von Terzaghi recollected Karl's special interest at that time: "Much later, when he was enrolled in a high school run under (protestant) church auspices, he wrote a paper highly sympathetic to the Darwinian theory of evolution. As a result, he came close to being expelled."

His grades in different fields in the school year 1898/99 were not superb with the exception of natural sciences which he liked so much. Karl graduated at the age of seventeen (Matura). He passed the final examination, however, already with honors.

1.2 "Sturm und Drang"

With his graduation Karl's childhood had come to an end. A completely new decisive period in his life began, a period which was very important for his whole life, development, and career and which was marked by a search for his attitude to life accompanied by errors, confusion, and insights as well as the search for the right profession. This process of searching for his true self lasted several years. In this time most of his later characteristics became apparent and his view of life was formed (Sturm und Drang).

Most influential in his personal life during the next years was his grandfather, Karl Eberle. He took him on three major educational trips: to Switzerland, Germany, and Italy.

Karl's descriptions of the tours read, in the main parts, as a travelogue with a detailed listing of all objects of interest. Little was said about his personal life and about his relation to his grandfather during his travelling. However, in the description of landscapes, Karl's talent became apparent, namely that of laying down, in written form, his observations of the Earth's surfaces, which he did so masterfully later for geological problems.

At first, however, in July, 1900, Karl started alone for a small visit to Zell am See near the grandiose Hohe Tauern with several peaks more than 3000 meters high. After a very rainy ride he arrived at Zell am See on Tuesday, July 10, where he rented a room. His objective was to explore the beautiful landscape close to Zell am See, namely the magnificent surroundings of Kaprun and Krimml.

Two days later, he started for his first excursion; he left his room early in the morning. The path led him endlessly through a meadow which became very marshy close to the lake. On a hill he saw to the left side the remains of a castle, and to the right, slate-quarries. He passed the old castle Kaprun, and entered and hiked through the Kaprun valley. His first station was Kesselfalls Alpenhaus where he had breakfast. After a strenuous hike he arrived finally at Wasserfallboden. Here, in the valley, the timber-line was already reached. Thereafter he went to a glacier which was covered by new-fallen snow. In the late afternoon he returned to Kesselfall where a bus brought him back to Zell am See. On Friday, July 13, Karl rode under a cloudy sky from Zell am See to Krimml. During the ride with the local narrow-gauge railway he enjoyed the beautiful views into the cross-valleys of the Hohen Tauern, in particular, into the marvellous Habach valley and the Sulzbach valley with the three and a half thousand meter high Großvenediger. He arrived at Krimml after a comfortable three hour ride and went to the first two falls, where he was surprised by lightning. During the next day, he hiked on a steep and stony path to the waterfalls of Krimml. "The forest is wild-romantic, to the right and the left of the path, mighty, glaring yellow and red rocks, covered with braids. At times one branched off the pass to viewpoints." After the third waterfall Karl entered the real Krimml valley. After a two and half hour hike he arrived at the foot of the Tauern and he enjoyed the grand view of the high mountain Reichenspitze and after a while the view of a mighty ice-field. After climbing and hiking through a marvellous world of rock- and ice-fields Karl, finally, arrived after passing grain-fields and farms, at Kasern where he stayed over night. On Sunday, July 15, 1900, Karl was back in Graz.

It is amazing how precisely Karl von Terzaghi described, already at that early age, his observations in nature. Following every detail he designed an instructive overview of landmarks, prodigies of nature and of geological phenomena. This talent made it later easy for him to formulate real geological and engineering problems which he published in numerous journals. Today, it is still enjoyable for experts to read his publications.

Karl started the first educational tour together with his grandfather to Switzerland at the very beginning of August, 1900. They travelled by train via Innsbruck to Lindau. By ship they arrived at Bregenz and Konstanz on August 4 and August 5, 1900. In Konstanz Karl visited all important objects of interest. In particular, he was interested in the old Roman-Gothic cathedral. In the late afternoon Karl and his grandfather rode by train via Schaffhausen to Zurich. They enjoyed the picturesque landscape with the island Reichenau, vineyards and the famous cataract in Schaffhausen. After breakfast, on August 6, 1900, they took a walk along the banks of the Zurich lake and in the afternoon they entered the electrical rack-railway which brought them to the well-known Polytechnikum. Karl admired the splendid statues in the vestibule, the old reliefs and the busts of former professors in the aisles.

The next station was Luzern. During a walk they got to know the originality of this town. The two wooden bridges with roofs and with unusual presentations of motives from Luzern's history appeared very strange to Karl. He was impressed, however, only by the splendid banks of the Vierwaldstätter Lake. In the afternoon they made an excursion by boat to Küssnacht, well-known from Schiller's stage play *Wilhelm Tell.* On Friday, August 10, 1900, they left Luzern and hurried to Bern. On the way they admired Interlaken's beautiful wood carvings and the marvellous view on the Berner Alps with Jungfrau and Eiger. In Bern they rented a room in the *Löwen* and took a walk through the peculiar old town with cellars emerging to the streets. The dinner was superb and on the following walk they had the opportunity to have a look at the splendid lighting system. Both ended the evening with a beer in a small garden. The next day was again filled with sight-seeing. First, they visited the magnificent late-Gothic cathedral, a masterpiece of architecture. Karl entered the tower and enjoyed the nice view of the Platte, a square-bastion – like a spur – located in the inner part of a loop of the river Aare, 40 meters above the river with the parliament building, town hall, university, and museums.

After lunch they prepared for the departure to Geneva with the express-train which was so completely crowded that it was hard to get a seat. They did not see much of the landscape. However, the first look at the Geneva-Lake and the Mont Blanc compensated for the stressful journey. In the evening they rented a private room and took a walk along the "Rue de Mont Blanc, the most beautiful street of Geneva" where they took their dinner in a restaurant.

On Monday August 13, 1900, they visited the Rhone banks, the Île de Rosseau and via the Pont de Monblanc the marvellous park on the right lake bank. While his grandfather was waiting in the park, Karl entered the tower of the old cathedral having a beautiful view. "The cathedral is with woodcarvings and monuments a building of interest of first rank." After lunch Karl visited the natural history museum, "for me one of the best objects of interest." Karl described exhaustively all the objects in the museum. After dinner Karl went with his grandfather to the Cafe *La Nord* in order to attend a string orchestra concert. On the next day, after breakfast they entered the steamer *Swiss* to go boating for 10 hours on the Geneva Lake. Having beautiful weather, Karl enjoyed the boat trip so much that he wrote an extra article on this boat outing with the title "La tour du Lac."

On Wednesday they left Geneva. After an interesting ride along the lake bank they arrived at Lausanne and further via Yverdon along the banks of the Neuchâtel-Lake to Neuchâtel and stayed there for one night. On the next day they rode by train to Biel. After having lunch they took a walk through the "town of the future", a really new town "because only a small kernel is old" and apart from that all was mostly new buildings.

After the sight-seeing tour they departed for Solothurn. Already at first sight the town made a favorable impression and they visited all the important objects of interest over the next two days. Then they rode to Aarau where they stayed only for one night because the town did not make a good impression on Karl and Karl's grandfather.

Via Zurich and Bülach they arrived at Schaffhausen on Saturday, August 18, 1900. They visited several points of interest on Saturday and Sunday and left Schaffhausen on the next day. Via Konstanz and St. Magarethen they entered Austria and rode to Landeck leaving Bregenz and Feldkirch behind them. In Landeck they stayed at the *Schwarze Adler*. After dinner Karl walked to the ruin Landeck and then through the Inn-Valley. On the way back he had a nice view of the Parseier-Peak. In the hotel he met his grandfather in company of an acquaintance from Graz to his great joy and surprise. On Wednesday, August 22, 1900, they left Landeck and rode back via Innsbruck to Graz.

The first major educational trip to Switzerland brought a lot of experience for Karl von Terzaghi, experience which was important for his further development. In particular, he saw many objects of interest, museums and the impressive Swiss landscape with the magnificent Alps. It is apparent that Karl was mostly interested in the description of the landscape, and not much in the history of Switzerland and its social and cultural life. He not only described the picturesque landscape in his diary, he also drew pencil drawings, some very nice and skillful, others, however, childish. He also tried to paint water-colors. However, he had to state that he had no talent for this kind of art; thus he chose only pencil drawings as a hobby which he mainly carried on during his vacations.

His relationship to his grandfather seems to have been a little indifferent. In no place in his diary did he mention discussions with him or other events related to him.

In notes, written on Saturday, September 14, 1900, he drew a summary of his trip to Switzerland. "During the Switzerland-trip from August 2 to August 23 I enriched my geographical knowledge and knowledge of human nature in a very considerable manner."

In short trips at the end of August and the beginning of September, 1900, he became acquainted with the Dolomites. He rode to St. Lorenzen where he immediately took a walk along a small run. He enjoyed the marvellous view of the numerous wild romantic mountain peaks.

After his departure from St. Vigil Karl hiked through the bizarre Dolomites, through canyons, along abysses on rock- and ice-fields, and lime plateaus. He described the geological formations extensively on eight pages in his diary.

After several days of hiking and climbing it was time to end the tour. "Now I hiked on the road further to Welsberg (Monguelfo, Italy), with a heavy heart taking leave of the Dolomites and used the eight-hour-train for the ride back."

During this time Karl von Terzaghi not only discovered the Alps but also tried to adjust to himself and to his environment. "At this stage I have left a Realschule-education (Secondary school) of seven years behind me and I feel the drive to prove which changes in my personality have occurred partly by my own will, partly due to outside influence. Once I was weak. I was sinking in the morass of many vices and I have to repent numerous slips. However, the drive for discovery has lived in me forever; I have recognized that happiness, character and drive are embedded only in complete self-control, ...

I can state to my pride that I have tried to improve my scientific findings. This effort has led me to doubt religion and materialism; in contrast to this I have heard about spiritism."

Despite all his impressive travels his vacations won their greatest significance through the careful study of Darwin's theory and in particular, through the fact that he became acquainted with spiritual ideas. Through this contact an important change occurred in his philosophy of life. Materialism, which he had payed homage, was recognized as incomplete. Religion, which he was very superficially familiar with, and which he esteemed little, he now recognized as a philosophical truth. He desired to build further on this ethical basis.

In the fall of 1900 Karl enrolled at the Technische Hochschule in Graz, just next door to his home in the Rechbauerstraße. The Technische Hochschule Graz had emerged from the Joanneum founded by Erzherzog Johann in 1811. In 1865/66 the technical school was raised to the *Technische Hochschule*

Technische Hochschule in Graz

Technische Hochschule in Graz and grandfather's house

with few faculties: engineering, mechanical engineering, chemical technology, agriculture and forestry.

In 1878 the Staatsprüfungen (state examinations) were introduced and in 1901 the Technische Hochschule received the Promotionsrecht (right of graduation). It was the wish of his family that Karl should study mechanical engineering for the next four years. However, this special field did not appeal to him at all and he was only occasionally seen in the lecture halls of the University. In exchange, he enjoyed the academic liberties in a rather desolate manner. His life as a student was wild and turbulent; he joined the bellicose fraternity *Vandalia* and spent most of his time drinking, rioting and duelling.

At the beginning of August, 1901, Karl von Terzaghi started his second educational trip together with his grandfather Karl Eberle – this time to Germany in order to visit the culturally most important places in the neighbor-state of Austria. They left Graz on August 5 and arrived at Salzburg in the afternoon. It was not easy to get a hotel room because a music-festival had just begun. On the next day they rode to Munich. In the afternoon Karl took a walk through the center of Munich and was impressed by its many magnificent buildings. During the next morning he visited the *Alte Pinakothek*, in which many of Munich's marvellous art treasures are collected. Karl regretted that he could not appreciate the works of art in the right way due to his lack of education in art. Two landscape paintings attracted, of course, his greatest attention. By accident, he saw some Böcklin reproductions. His paintings had an irresistible fascination for Karl. They awoke different emotions in him which he could not explain and found rationally. He recognized that he had to fill this gap in his education. On the next day he went with his grandfather to the *Neue Pinakothek*. The impression was almost more lasting than that of the *Alte Pinakothek*. He again admired the landscape paintings

He was a little disappointed by the *Ruhmeshalle* (Hall of Fame) with the *Bavaria*. On his last walk he saw the *Maximilianeum*. A broad and grand avenue led him to the object of interest, with its main memorials to great German scholars. He looked from the Isar-Bridge into the park adjoining the building, a real work of art. He took a seat on a bench for a while and admired the bushes, illuminated by the evening sun, with the red-lit Maximilianeum gleaming through.

At Maximiliansplatz they crossed the trolley lines. Karl was lost in thought, thinking of all the impressions he had gained. Suddenly, he heard a pitiful crying behind him. He looked back and was for the first moment paralyzed with terror. Karl saw his grandfather lying on the trolley lines, being dragged away by the trolley. This was a terrible moment. At last, the trolley stopped and to Karl's great surprise and joy his grandfather got up. Some passers-by led him into a house. Karl inspected his grandfather's injuries. However, fortunately, there were only some skin abrasions and swellings, which were quickly bandaged.

Karl as a young student, 1900.

The plan to leave Munich on the next day was of course in a sad manner cancelled. Karl and his grandfather spent the next day (August 9) involuntarily in Munich. The first trouble was soon overcome and in the afternoon Karl took a look around in the magnificent *Englischer Garten* (English Garden).

Karl and his grandfather continued their tour on August 10, to Nuremberg, the old town with battlements, turrets, and dark narrow lanes – recalling the Middle Ages. On the next day, Karl took a walk along the *Graben* and he really got the impression of an ancient castle. 34 years later Karl would return to Nuremberg on business.

Karl and his grandfather visited the next points of interest in Eisenach on August 12, 1901. Karl enjoyed the lovely Thuringia forest on the next day. Already at four o'clock a.m. he started his trip through the forest solitude; he admired the Warthburg on his hike and he recalled the place of this landmark in German history, in particular, its importance for the fraternities. In his thoughts he called on the students to keep the joy of songs and weapons and to conduct the bright *Schläger* (rapier) with honor.

In the evening they rode to Leipzig. However, Karl did not like this town; he felt it was too plain. Therefore, Karl and his grandfather headed on further to Berlin in the evening. The next morning he started his walk through Berlin in the famous street *Unter den Linden.* His expectations were not let down. The broad street was encircled by splendid buildings, royal castles and lovely parks. On one dignified end was the *Brandenburger Tor* and on the other side the Schloßbrücke. Beyond the bridge he saw the highlight, namely the *Lustgarten.* Moreover, he admired the cathedral and the museums on the museum island. Furthermore, in the afternoon he visited the zoo, and in the evening, the observatory.

On August 16, 1901, in the afternoon, Karl and his grandfather made a trip to Potsdam in order to view the castle Sanssouci. Although he was not in the right mood, he enjoyed the stylish beauty of this castle located in an expanded, marvellous park decorated with statues, temples and fountains.

Two days later, Karl and his grandfather made their way to Hamburg. They stayed there in a splendid hotel which was located at the Alster. In the afternoon they took a walk through the city. Karl was very pleased with this interesting and original town: "a German Venice." On the next day they boarded a boat and visited the harbor, one of the most famous harbors in the world.

On August 21, Karl did his private correspondence and after he had completed this he went with his grandfather to the railway station and rode to Bremen, to the other main harbor on the North-Sea. The ride was a comfortable one, partly through forests and partly over waste heath-land. "The most interesting place in Bremen is without doubt the market. Encircled by splendid old houses, patrician and guild houses, the stock exchange, and, in particular, the city hall. In front of the old landmark of jurisdiction, the mysterious Roland. The Gustav Adolf and the Wilhelm monuments are located here. The interior of the city hall is in accordance with the outward appearance, splendid and antique. On the ceiling hang some small sailing-boats, the walls are covered with carvings and frescos. Also the cellar, a real drinking hole, is quite historic. Huge barrels are set up on the walls for thirsty souls." In the evening Karl proposed to already travel to Cologne on the next day. Karl's grandfather agreed immediately, as they had seen all points of interest. The ride to Cologne took seven hours. In Wanne (Herne) they entered the Ruhr area with all its heavy industry, plants and coal mines and the "nice small houses of workers." In Cologne they crossed the mighty river Rhine and arrived at Cologne's main railway station. Soon they found a room in the *Ewige Lampe.*

On the next morning, August 23, 1901, they visited, first, the Cologne Cathedral. "In front of our eyes the mighty Cologne Cathedral, a colossus, rises; the dimensions are beyond any imagination. We entered the interior through a memorable large portal and were astonished at the simple greatness, at the modest harmony of this house of God. To the right and to the left the columns, mighty, like the trunks of a very old forest, strive upwards, so that one can hardly follow them and end up there in mighty vaults. The faces of the apostles and martyrs looked seriously from the huge glass windows ...; each of the monumental sections breathes the idea of God and the withdrawal from the bustle of the world. One could not erect a more dignified monument for the God of the Christian world."

The trolley took them through the city. Karl's interest in the buildings was not great; his thoughts were still with the splendid Cologne Cathedral. In the afternoon he went again to the Cathedral. A doorman led him via a winding stair to a gallery where he stood in the middle of the architectural sections admiring all the details of the construction.

In the tower he met a young man from Berlin. He enjoyed the time with him and was glad to exchange his ideas with a person who felt like he did.

On August 24, Karl and his grandfather made their way to Mainz via the river Rhine on the steam boat *Auguste Victoria.* This trip is some of the most beautiful boating in the picturesque landscape of Middle Germany. Soon they left the flat land. To the left they saw the Siebengebirge with the castle *Drachenfels*; moreover on the peaks of the mountain the first castles, silent witnesses of the long German history looked down upon the Rhine. In Koblenz the boat brought them into the real Rhine Valley with the most spectacular views of the river banks, mountains with wine, castles and a mysterious charm. Karl von Terzaghi described the boat trip on the Rhine with at times florid language "One is not forced to think of a wild, romantic canyon. No. However, the woody heights, picturesque interrupted by vineyards and rocks. In the basins the most lovely towns thinkable, are still in their Middle Age garb ... That all combined with the venerable wreath of legends, which are tied up at every corner and every place of this marvellous region has, with the great historical past, an indescribable magic for every receptive heart." At the Loreley and the Rhine Rock these feelings reached their climax. "The prosaic spectator would say a slate-rock which projects into the Rhine. I say it again: The marvellous surroundings, the green stream, in the distance, the lovely St. Goar, and the old castles, in front of me the water rushing down grey rock is rising sphinx-like, the silent witness of the past of this place. I was on the verge of tears. All the feelings, which many inspired hearts of poets were able to feel in this place, passed my spiritual eye; I felt the whole solemn mood. My thoughts roamed the past to that man, on whose side I spent the most beautiful hours and whose memory I will also always keep in such moments" (Probably the man he met in Cologne is meant, the author). "The Loreley has passed out of our sight and here the incomparable Rhine Rock emerges on the deep bank from the green beech forest. The charm and the fascination of this region is simply indescribable. The idea, as to how one can imagine the castle in the old heroic legends, seems to come true. From the stream the hill emerges steeply surrounded by marvellous beeches. Rugged, adventurous ribs of rock and sharp points interrupt the green and here on the rugged tooth-like rock the castle rises as picturesque as possible; on the side a noble Gothic chapel. I will describe nothing in detail; it is impossible for me.

Now, we have left Bingen. I look back to the splendid *Niederwalddenkmal*[5]; the *Siegesstatue* contrasts with the red-hot sunset glow. In silent

[5] Monument

majesty it makes known widely the glory and the greatness of the German nation; the monument is reflected in Germany's most famous stream. Not yet have we arrived in Mainz and already now the mountains contrast with the black of the night sky. The moon is reflected in the marvellous stream which seems as if covered with silver light. Meditating I am looking from the railing to the waves and letting pass before my eyes all impressions which I have felt on the unique Rhine today."

On Saturday, August 25, 1901, Karl and his grandfather spent the whole day in Frankfurt. Karl was not in a good mood. He did not like Frankfurt very much and also his relation to his grandfather was not the best in this moment. "My grandfather is a kind, good man, full of goodwill, but he is already old and I am still a young man. Both are not well matched. He, already detached, sees the world with cold eyes, and I, still full of enthusiasm and fantasy; it becomes unbearable, to see always a person around me who does not share my feelings. It seems to me that I cannot bear this state any longer. Wild indignation varies with complete indifference; I have felt this state often already, now it is most vehement. Oh, how I shall breathe again, on the mountain, there is freedom! What valuable enjoyment this trip through the German provinces would have at the side of a young man, full of enthusiasm and emotion, like me. Of course, at the beginning nothing would please me and as before mentioned today the feeling of constraint, of discontent, was stronger than ever. At last gradually the shadows vanish and I unwind slowly."

A highlight of Karl's Germany trip was Heidelberg. "Marvellous Heidelberg. There is hardly a second place on earth to be found which can be compared with the beauty, with the poetry of the Heidelberg castle. Here nature and the work of man are combined in a perfect harmony, which defies every description. At this place nature seems to have spent all its charming attractions. The reality exceeds the fantasies of a romantic, splendid castle. I am sitting here in the deep moat, all round rise very old ivy clad beeches and linden trees and only some lonely sunrays get lost in the shady ground. A mysterious rush goes through the forest ceiling, on the ground a source ripples talkatively. While I am writing in the moat, I notice a broken winding staircase, uncovered tunnels and vaults. All things carry the traces of time; climbing plants cover the sad remains."

Then, Karl left the moat and walked to a marvellous forest, to the large terraces where he viewed the old, red castle, and Heidelberg town divided by the river Neckar. After visiting other points of interest in the castle he said good-bye to this splendid place. In the morning of August 28, Karl and his grandfather left Heidelberg and headed for Ulm. However, the weather was so bad that they decided, after a short stay, to ride on further to Munich and to Salzburg, where they had lunch. Karl was very happy when they came to the Alps. "... I must state that our fatherland, the Alps, are beautiful, too. So beautiful, so marvellous also the nature; there is only one thing missing:

the great past, the memory – thus the working of a strong nation, that past which ensnares with such ever lasting magic the Rhine, the Black Forest, the Thuringa Forest and other regions."

As usual, the Eberle/von Terzaghi family spent their summer vacation in the countryside. This year the family had chosen the Grundlsee. Karl and his grandfather, therefore, rode to this marvellous place. Their arrival came as a surprise. Karl was very happy to be back after this long journey and to see his family again.

During the next days he recovered from the exhausting trip. With his sister Ella he went boating on a lake. Due to her kind and gentle character he enjoyed her company. She told him that she had had a boring time during his absence and that she was happy to see him again. Ella told him about some bad occurrences of the last month. For example, that his grandmother had had the intention of opening Karl's letters in order to get some information about his plans in order to spoil them. He was bothered that his grandmother played a false game. In contrast, he got to know his grandfather's best side during the trip. "He is indeed well-meaning and gives me advice in an always reasonable manner ... He is in most cases a very sensible adviser although due to his age his views are sometimes clouded."

During this time Karl von Terzaghi tried more and more to put his life in order. As mentioned earlier he was not bothered by his study at the Technische Hochschule where he visited the lectures of the professors very seldom due to the fact that he was not very interested in mechanical engineering which he had chosen to study one year before. However, his main points of concern were on the one hand the chaos in his thoughts and emotional life and on the other hand the conflict between his intellect and his emotions. On September 1, 1901, he wrote in his diary: "Today there is another day in which I cannot bring, much as I should like to, order into the chaos of the thoughts, and such days are dangerous for me. In my brain lie so many and so different ideas and thoughts and power in confusion that I am not surprised." Also, the chaos in his emotional life disturbed him. As already indicated earlier his academic and, in particular, his fraternity life was filled with exuberant and wild extravagances. After such "excesses" he criticized his own behavior heavily and promised to change his life. However, after a while he fell back into his chaotic life-style although he made great efforts to understand his characteristics, to master his passions, and to strengthen his intellectual abilities by reading books on history, art and philosophy. In order to bring his thoughts into order, he read a book *History of Painting*. On this occasion he had the following idea: "Arts seems to be the same for feeling and spirit as science for the intellect ... What is spirit? I cannot answer, I have only a hunch, but I cannot explain. Intellect and spirit seem to be two of the driving forces of the human mind, perhaps there exist only these two, both have their source in the causal requirement. The intellect requires the recognition of the logical continuity of occurrences, the explanation. The mind receives impressions, feels thereby beauty, goodness, and feels more or less the impetus to depict the perception in colors, words or tones. The mind demands therefore a certain passive realization in contrast to the activity of the intellect which wants to enter the inner part. While it puts up with this to luxuriate in moods and feelings. Hand in hand with the refinement of taste, the realization of beauty etc., goes also the perfection of art. The different steps of art are the real reflection of the Zeitgeist (Spirit of the Times) dealing always with that, which the contemporary person considers as the most beautiful, eminent, venerable."

Moreover, he was occupied by poems, philosophy, evolutionary theory, and literature. On September 5, 1901, he read Lenau's poems and he was impressed by the melancholy verses, deeply penetrating into the soul. The bounded language had an enormous, mysterious power over him. He listed in his diary many works of humanist discipline: Kant's and Schopenhauer's works, writings of Spinoza, Leibniz, and Kleist, poems of Schiller, Goethe, and Uhland, but also works of natural scientists, namely Pascal and Darwin.

He started again to read the writings of the philosopher Gustav Theodor Fechner, physicist and philosopher (1801 – 1887). In his philosophy influenced by the Late Romantic, he was an exponent of a speculative theism on the basis of pantheism. Karl admired Fechner's ideal enthusiastically. "*The old God is still alive* one should cry out when one catches sight of Fechner's pantheistic universe penetrated with spirit ... In this teaching of Fechner the whole sunny, optimistic picture of the world of classical Hellenism seems to revive. Nothing more of the dark, pessimistic thought, the world seems to be turned into an overhappy, bright paradise. The human being is no longer an unblessed protein and a clot of protoplasm, but rather a systematic being provided with different spiritual abilities. From all this I can see that the aesthetic and also the practical value of such a picture of the world is inestimable, the effort will hopefully pay off."

On the next day, September 8, 1901, Karl von Terzaghi continued his profound reflections on Fechner's philosophy. "Like a warming flood of light Fechner's teaching penetrates my heart. I am already beginning to catch the notable, morally invaluable, thoughts of this prophet. Full of gentleness and warmth he preaches his wisdom; it appears like a transfigured Christianity. Here I have found, finally, the base, where I, up to now, aim- and planlessly wandered, finally a floe in the stormy water."

By September 6, 1901, Karl's grandparents had already departed towards Graz and now he enjoyed the airy balcony and the large, light room which he could now use alone. He went boating in the morning and drew some sketches in his sketch block. During the following days he really relaxed with boating and hiking, and sometimes he took a nap. His mother and his sister accompanied him from time to time when he went to Aussee. During the walks and at home his thoughts soared to philosophical questions and, in particular, to the works of Fechtner. Moreover, he reflected on Lindenberg's[6] *Berlin*, and on this occasion recollections of the marvellous Sanssouci, and its admirable builder Frederic the Great came to his mind. "This residence, this surrounding and society, in which Frederic stayed, and the fine spirit which was inherent in him, led this place on earth to appear as a paradise on earth at that time. I admire the excellent, fine and noble spirit of this man. He did not impress by his power and the scope of his knowledge, but by his nobility, the veritable royalty, which distinguishes his thoughts."

"What a difference between the distinct great minds: the modest Kepler, the fine Leibniz, the bulky Wallenstein and the noble Frederic, each of them a distinct personality; I remember the comparison with the peaks of a mighty mountain chain, which rises high into the dark sky. All different and nevertheless magnificent in their kind." The enthusiastic admiration of Frederic the Great is only understandable if one considers the age of Karl von Terzaghi at that time. Today, we know that the francophile Frederic knew very well how to use his power and that not all of his thoughts and actions were imprinted with pure noble-mindedness.

On the same day he continued his reflections on Frederic the Great. "Not only his actions and his words I should observe but, in particular, his thoughts which are the source of both, aforementioned. This point I have failed up to now. I have to come to learn again today what poisoning action, in particular, lascivious thoughts have. The whole soul is put into a state of pronounced listlessnes, loosening. A clear collection of the spirit, conduction of the thoughts to good and beauty, this is the only right thing. As far as this worldly wisdom is concerned, Emerson[7] seems to be the best leader and advisor. My maxim must be: All, that whole of literature, the best minds, can provide I will consume, with a clear view looking through and sift in order to build ..., with own ideas, my heaven."

On September 20, the weather was superb and he decided to hike through his beloved mountain range, the Dachstein area. He packed his rucksack immediately with all the equipment he needed for the tour. At one o'clock he had already reached his aim. He spent several days in the Dachstein area (Hallstadt, Simony-Hütte) with hiking, climbing – and sketching, his favorite hobby in recent time. After the very strenuous tour he came back to the small vacation house. Obviously he was so exhausted that his sister did not recognize him at first.

His sister had struggled during his absence with her mother because of Ella's interminable crochet work. "I do not know, is she not able to find another hobby. She will become totally stupid and unsociable. It is strange that she does not like any sociability and therefore she is popular nowhere. I will do my very best to bring Ella in contact with the outside world. In

[6] Paul Lindenberg, a 19th century writer, authored descriptions of towns and landscapes.

[7] Ralph Waldo Emerson, 1803 – 1882, was an American essayist and philosopher.

most points she is the direct contrast to me. And actually she is such a good, seeming naive and dear girl."

Karl's mother thought always about his hiking and climbing in the Dachstein area and she could not calm down about the risk her son took.

At the end of September, Karl got a little home sick. "I cannot deny: it is wonderful on Grundlsee, friendly the green forests are still shining, however, the loneliness, the inactivity and the absence of outer diversion forced me to depart."

In the next days he developed many wild sexual fantasies and his desire to visit the brothel in Graz became very intense. However, he had a bad conscience. "From the standpoint of nature, custom and morality there is actually nothing more abominable than a woman who prostitutes herself for money in public. This thought alone should keep me away from these whores. However, the lust pushed me back again. Which way should I adopt? The seductive girls jiggle in front of my eyes showing their feminine charms totally frankly, the well-developed breasts and piquant black stockings with the delicate patent leather shoes, the pleasantly smelling stairs and warm beds ... What, you rascal: Here your mouth speaks of noble Germanity, of free science, of fair mindedness and pure thoughts and nevertheless you want, in shameful embrace with a prostitute, to sneer at any moral. You are preaching about the ideal of a healthy and beautiful body, however, in a criminal embrace you suck in the dreadful poison, which changes the human being into a living but putrefying corpse. No, I should not stand this conflict in my nature, get away with all of these shameful and so to speak more or less prostituted thoughts ... I will turn my thoughts to the pure, beautiful hymn,

The members of Vandalia (Karl second left in back row) about 1901

lift myself in a brutal way from the mud. In this place I swear, and I vow, and I will keep the oath, to stay away from this vice from this moment on."

On his last day at the lovely Grundlsee he reflected again on his life. "I have conflicts with my relatives almost constantly, I must go along with the circles of acquaintances, I must fearlessly handle the foils and must always bring to bear my person." Moreover, Karl referred to the philosopher Emerson: "it is easy in the world to live according to the world's will or in solitude according to one's own will. 'Only that man who knows to keep his originality in the world is the true man' says Emerson and according to this principle I lived when I enrolled at the Hochschule."

When he was back in Graz, Karl continued his exciting, restless and eventful life. On November 23, 1901, he was already at the *Kneipe* at 8 a.m., drinking steadily and throwing dice enthusiastically with his beloved friends. He was already in a good mood at 11:30 a.m. when suddenly two other friends entered the room with much noise and asked them to join them and to ride to Leoben, to the Ledersprung[8] (Leather Jump). Karl agreed immediately, ran back home, asked his mother for 10 Kronen ("bummed 10 Kronen from my mother"), pulled on his black saloon suit and just 15 minutes later the four Vandals were at the railway station. The ride went by very fast with jokes and merry talks. In Bruck they had a snack and in Leoben they were expected by some members of the fraternity. The splendidly furnished Kneipe, the nice reception and the good beer impressed Karl very much, so that he drank a lot of the stuff. Already at 7:45 p.m. the *Fuchsmajor* stopped Karl's drinking. Later, he was on his way to the hotel Post, where the Ledersprung was to occur. The numerous guests had their seats on a horseshoe-shaped desk and the first-semester students of the mining academy were sitting at an isolated table. The Biersperre (beer blockade) was stopped and Karl again drank many glasses of beer, so many that he passed out. His friend brought him to his apartment and threw him into his bed, so roughly that Karl hurt his nose seriously.

The next day was again filled with beer and wine drinking, and on November 25, Karl was back in Graz and a carriage (Zweispänner) brought him to a friend's apartment where he relaxed from the drinking trip.

---

[8] The Ledersprung was the main event in student life in Leoben at the beginning of the winter-semester. The students wore mostly a miner's overall and the members of the fraternity also wore a band and cap of the fraternity. The members went to the beer-barrel and formed a lane. Then, the Fuchsmajore (fox-majors) proceeded with their Füchsen (foxes) in a single file to the barrel, and also the Finken (finches) who were not members of a fraternity. Then the Ledersprung followed for every young student. He had to climb the beer-barrel with a glass of beer in his right hand. The president asked him for his name, his standing (mining or metallurgy), homeland and his motto, and he finally said: *So empty your glass and spring into your standing!* Close to the Füchsen stood two guests of honor holding a leather apron. After drinking all the beer the student jumped over the leather "into his standing".

However, the abstinence from beer did not last very long. Already. on November 28, Karl was sitting in the Kneipe and had the idea to invite his friends to arrange a party in his room at home (Budenabend). He returned to his home, arranged everything and soon five friends came with 1.5 liters of schnaps. The corps brothers were drinking and smoking a lot, the mood getting better and better; in his bed stood the rocking-chair and lay bottles and his writing desk was a puddle. Finally, Karl stopped the party. However, they continued it in a restaurant and Karl's friend took him to his apartment at 3:30 a.m. in the morning. The next weeks ran by in the same way. For example, in the middle of December a Corps member who had become *Fuchsmajor* invited five Füchse (foxes) to a drinking evening in the 2nd floor of a house in the Schmiedegasse at 7 p.m. Within a relatively short time the six drank hundred of bottles of beer of different sorts. The consequence was the complete soiling of the floor and the court-yard with their excrement. Even the first floor was not spared. Around midnight they left the house, making a lot of noise, and one of Karl's friends was arrested because of disturbance and extinction of lamps. While his friends struggled with the policemen, Karl wiped out a red lamp. Shortly after that they rushed off to the brothel.

Karl spent New Years Eve 1901/1902 in the *Kneipe* with drinking and discussing with his cronies. However, shortly after midnight he left his friends and went home. He wanted to let the previous year pass in review and to let his imagination wander into the future. (The relatively calm celebration of New Years Eve became almost tradition in Karl von Terzaghi's further life). He was, in general, pleased with the developments of the last year. However, one request came from his heart: "As 1901 was a year of tranquil thinking, of wavering and of choosing, so 1902 should become for me a year of action."

However, despite his good intentions he spent again much time with his friends drinking and thinking up unusual pranks. One night he and a friend made a lot of noise in the city park. When a policeman appeared they climbed a tree and abused him. Moreover, they relieved themselves above the policeman. Finally, in the course of the incident, Karl and his friend were arrested and the newspapers reported the story on the front page under headlines such as *Strange birds in a Tree in the City Park.* The amused judge in the court acquitted both, "but the police department was furious. They appealed to a higher court and won the case. We were thrust into jail and received additional punishment on account of 'disorderly conduct in jail', because we had bribed the jail and turned our sojourn into a noisy and jolly party." Moreover, "the incident was brought to the attention of the faculty, and since we already had a bad reputation and a rather voluminous police record the consensus of opinion was that we were ripe for expulsion." However, there had only been three expulsions in the entire history of the Technische Hochschule of Graz. The first one contributed an important invention to electrical technology; the second one constructed a steam turbine; and the third one became a famous church architect in Germany. Obviously the Technische Hochschule had expelled the wrong people, and the professors were beginning to dawn on them that the young student Karl might develop in the same way as his predecessors. It was Professor Wittenbauer[9], Karl's professor for mechanics, who turned the tide, and who became his fatherly friend until Wittenbauer's death.

At the beginning of the new year he spent much time with different friends. The discussions between them were concerned with the right adjustment to life, in particular, to his fraternity. In the first draft of a letter to a friend from March 15, 1902, he specified his opinion of the fraternity. "The Corps is a company of young men, a safe retreat of real friendship. Everybody is responsible for every one else, up to the last drop of blood. This is the ideal of youth: to have friends, upon whom one can rely. We spend together the nicest hours, we drink, are enthusiastic and sing and shed our blood for our banner if this is necessary. However, where are the spiritual pleasures, the improvement of the human being? Where is this marvellous ocean of hopes, of supreme happiness? This should slowly go numb in me?" After, praising the fraternity he had to report to a Corpsphilister[10] in April, 1902, that they had much trouble in the fraternity. In the course of the winter semester several Corps students had to be dismissed. When they audited the books they found some financial mismanagement, which surprised them. He requested the Corpsphilister to donate money. At the end of his letter he reported that the inner Corps-conditions had been much improved. "As a Corps-Bursche and then as a Senior I succeeded in bringing the strictest discipline to the Corps." Moreover, Karl stated that the rough tone at the Kneipe, which was brought into the Corps by a member of the fraternity had

[9] Ferdinand Wittenbauer was born in Marburg on the Drau (Maribo) on February 18, 1857, as the son of a regimental medical officer. He attended the Oberrealschule in Graz and studied in the engineering faculty of the Technische Hochschule in Graz. In 1876/77 he served as a volunteer in the army for one year and was appointed lieutenant in Vienna. He finished his studies with the diploma examination at the Technische Hochschule in Graz in 1879. Then, he became assistant at the chair for road and bridge construction (1879-1882) and did his habilitation thesis in theoretical mechanics and graphical statics.

Wittenbauer had been on sabbatical leave at the universities in Berlin and Freiburg im Breisgau (Germany). In 1891 he took over the full professorship for mechanics and theoretical mechanical engineering. He was appointed *Rector* of the Technische Hochschule in Graz in 1911/12.

Wittenbauer is credited as the creator of graphical dynamics, a field which has disappeared completely in the modern era. He not only worked in his scientific field, but rather he was in addition a noted poet and playwright, e.g., *Der Narr von Nürnberg* (The Fool of Nuremberg) and *Der Privatdozent* (The University Lecturer). Moreover, he was very conservative and a racist like many other people in this part of old Austria in this time.

Ferdinand Wittenbauer died in Graz on February 16, 1922.

[10] A Corpsphilister is an alumnus (Alter Herr) of the Corps, whereby Corps denotes a special fraternity with idealistic goals.

totally vanished because this member was forbidden under severe penalty to use such a rough tone.

On May 16, 1902, he expressed, for the first time, some thoughts concerning his studies. "Now, I will concentrate on my studies. I will begin with mathematics."

On August 10, 1902, Karl spent for the time being his last day at Grundlsee where the family was again on vacation. The whole day he packed for his third and last educational trip with his grandfather, to Italy. He was a little afraid that he was not this time well enough prepared for this wonderful country. The only book about Italy which accompanied him was Goethe's *Italienische Reise*, "a pearl for his power of observation and mind, a brilliant example for all hikers." On the next day they started on their long trip. Karl was not in a good mood; the train roared through dense fog and sometimes rain splashed against the windows. After a long ride they arrived at Villach. The whole town did not impress Karl very much. Thus, he wondered about harmony, about pessimists and the architecture of some buildings in Villach.

He was much more impressed by Udine: "When I entered the main place of Udine today in the morning, a comfortable feeling came over me; this was again a view which revives the eyes. Refined lines, noble forms, no disharmony." Then Karl von Terzaghi described the town and its inhabitants. "The houses are narrow and high and not regular, ..." However, Karl was not bothered by the irregularities. "The labyrinth of narrow streets ... with palaces and small houses are inhabited by a carefree, always cheerful, extremely naive native crowd." In the center of the town Karl could see the old castle on a hill. His opinion of the Italians was not very high: "The Italians are, as it appears to me, not a very capable and vital nation anymore. They live, so to speak, under the protection and under the nimbus of their great ancestors, a harmless, carefree and joyful existence."

From Udine they rode through fruitful and lovely Lombardy, full of nice country houses and passed the fascinating lagoons of Venice with the great colour harmony.

On August 14, they stayed for a while in Padua, where the great Galileo had spent so many years. Karl described in detail the town, in particular, the Basilika and they arrived in Bologna, located at the foot of Alpenines, at sunset. The colour effects were so wonderful and marvelous that Karl was unable to describe this fascinating spectacle.

Florence, the jewel of Tuscany, was the next destination. On this day (August 15) Karl was very depressed. Moreover: the constant griping of his grandfather made his state worse. Nevertheless, Karl took a walk through the metropolis of arts. He was deeply impressed by the Palazzo Pitti, designed by Brunelleschi in 1440 for the family Pitti and later enlarged to the largest palazzo of Florence. It looked to him magnificent, unforgetable, though however, severe and depressing.

On the next day, August 16, Karl von Terzaghi visited the Uffizien which houses one of the most important collection of paintings. Besides a complete survey of Tuscanian art (from Giotto to Botticelli, Leonardo da Vinci and Michelangelo) the Uffizien also contains magnificent works of great painters from other Italian regions (Raffael, Tizian and Tintoretto) and German (Dürer, Cranach and Holbein)as well as Flemish and Dutch masters (Rubens, Rembrandt). The marvelous wall- and ceiling-paintings were worked out by Vasari and his students. Sculptures from Michelangelo, Giambologna and Verocchio embellished the rooms. Karl admired, in particular, the Laokoon group and two paintings by Rubens. He regretted that he could not see all the endless splendor due to their short stay in Florence. For the same reason he could take only a hurried look at the cathedral.

The next day's stop (August 18) was Rome, the legendary capital of Italy with its rich history. Karl stayed with his grandfather for three days in Rome, surely too short to see all the points of interest. Karl restricted himself to visiting the main points, namely the St. Peter's Cathedral, Forum Romanum, and the Colosseum, and he enjoyed Rome's atmosphere. He regretted very much that his stay was so short. "Hardly do I begin in Rome to awake and to look more freely with more understanding and already now I must leave. It is sad to have to part. Nearly everything remains undigestible. I have experienced only a glint of the greatness and the beauty. Rome has the effect of enobling when one keeps out all the vulgar things." At the end of his vacation in Rome, however, he expressed some optimistic thoughts. "The language of art becomes clearer and clearer to me in my mind. The breath-taking tragedy of Michelangelo's Doomsday, that crouched figure who desperately tears its hair, had a mighty impression on me. A great, heavy plan grows slowly in me. I will describe our contemporary situations in a work ..., will lay down to myself and others clearly the relations in which we stand to art and science, will explore the relation between mind and intellect, will put into words the danger, my hate against weakness and absentmindedness and will free myself from that urge with which we are all hereditarily tainted."

The southernmost town of Karl's trip to Italy was Naples. His vacation lasted from August 21 to August 26. Naples is located in a fantastically beautiful landscape on the deep blue gulf of the same name, with Vesuvius in the background. Guests are fascinated in this town by the colorful and noisy vitality in the streets. The narrow lanes with many, colorful shops and the washing hung between the houses, are very picturesque.

Karl von Terzaghi was depressed on the first day. It may be that he had got tired because he had already spent ten days in Italy and the tour was strenuous. On the next day he felt much better and he visited the Pompeiian museum. However, the museum was crowded; thus he got only a perfunctory impression. The rest of the morning he used to bathe in the Mediterranean Sea. In the afternoon he visited an aquarium together with his grandfather, where he could admire the colorful fish; he was pleased and left the aquarium. Outside he enjoyed a spectacular sun set.

On August 23, Karl and his grandfather went boating. The most interesting place for Karl was obviously Capri. He mentioned, in particular, the Blue Grotto, where the water shines a splendid sky blue. Unfortunately, he missed the Palace of Tiberius, because the time was too short. The ride back on the deep-blue colored sea was splendid. When they rode into the Gulf of Naples, the sun was already sinking and lit up the rocky coast.

The ascent of the Vesuvius on August 24 was a real adventure. He rode with the horse railroad to Resina and rented a horse. A guide and the owner of the horse accompanied him. The small town was crowded and it was not easy to pass the crowd. Finally, they arrived at the fruitful, sunny Compagna. The nice country houses lay idyllically in luxurious green. Soon the landscape changed; they arrived at the black lava stream. The rigid surface without slag wound in the most bizarre form through parts of the vineyards. Finally they came to a group of country houses overlooked by the seismic observatory. In a simple restaurant they ate a meager meal. The marvellous view of the Gulf and Capri from the terrace compensated for the simple food.

Thereafter they again mounted the horses and rode on a small path to the ash cone. The landscape became more and more deserted. They left the lava fields and walked through the loose deep ash. The horses could not stand this and therefore they tied them up on a mighty block, and accompanied by the guide, Karl started the ascent of the ash cone. Finally, after strenuous climbing they approached the peak. Sulphur vapour broke through fissures and caves and the loose rocks were colored yellow like. Then they arrived at the "jaws of hell." From the bottom steam arose steadily. Soon the atmosphere became unbearable and the guide took Karl to the other side of the crater. From here they enjoyed the beautiful view of Monte Somma; then they returned to Resina.

On the next day (August 25) Karl and his grandfather brought their stay in Naples to an end with a tour of Pompeii; for Karl the visit exceeded all expectations. "In former times I did not hardly dare to foresee what I see here accomplished. Through a splendid garden the way leads to the town of ruins. Through a gate we enter a narrow lifeless lane and soon we arrive at the forum. The place is regularly rectangular and almost bound by arcades. In the background the wonderful ruins of the Jovis temple rise, to the right and left we see public buildings and temples of other gods. Each building is of classical calmness and beauty. We leave the place and admire the private homes. The familiar design is of perfect beauty and expediency, the model of residential premises. How gladly I would like to give free rein to my imagination to picture the life in such a domestic temple.

The artistic frescos on the walls are representations out of the rich mythology or refer to the occupation or the inclination of the master of the house."

Karl promised to write a profound article on Pompeii. However it seems that he dropped his plan.

Karl and his grandfather left Naples on August 26, on a large steamer for Genoa. Karl was so impressed by the sea trip that he wrote an extra article in his diary. He admired the coastline and enjoyed the opulent dinner on deck. On the next morning the ship made a stop-over in Livorno where some passengers were brought to the town and freight was landed. On August 28, Karl and his grandfather arrived at Genoa. On the next day they rode to Milan, where they stayed for two days. The most interesting building for Karl was the cathedral, which he really admired. The first part of their trip home was pleasantly calm. They spent the last evening in Italy. "The separation is becoming hard for me. I esteem this marvelous country more and more. Everything here is more beautiful. Under the free sky spirit and nature sprout luxuriantly and perfectly. The works carry the stamp of bright sensuality and chosen love of life ... Farewell happy Italy, when and how will I see you again? I part happier and richer and stronger." On September 1, Karl was back at Grundlsee, back from his third and last educational trip which had introduced him into the marvellous mediterranean world.

During the next days he spent a lot of time thinking about his life. He developed different confused thoughts concerning his intellect and his mind. He described his philosophical thinking about his intellect in a special section in his diary with the title "Anatomy and Pathology of my Intellect." The main result of his thinking was: "The only thing the terrible hours brought to me, was the realization that the intellect is not sufficient to comprehend and to solve the basic problem of existence, that all attempts of metaphysics mean fruitless efforts. I am indebted this moment, in which I found this truth, to everything. Suddenly I recognized that mysterious double nature of all things, the puzzle of the world: nature and intellect are the same, only separated by our subjective feeling.

New life flows through my veins. I felt the God all present, which is in every molecule, in each plant, in each planet, in the ether. Gradually I understood the sensitive, mysterious language which the little plant, the tree, the ruin, the work of art whisper to the listening mind. Slowly, very slowly this realization here formed and grew and now I am there, a human being among human beings, deeply rooted in soil, looking forward to keen work, aim oriented creation. Other people unconsciously enjoy the beauties of nature, the blessings of arts, I have to struggle hard for all of this. The whole realm of the gods of the Greeks comes to life in me again in a new, noble, true fashion.

That, which once was preached by great intellectual giants in wise symbols, in poetic forms, I see it in front of me, pure, clear, beautiful.

Separated, both powers take up their ways: intellect and mind, usefulness and beauty. In divinity both are tied up in complete harmony." Finally, Karl von Terzaghi summarized the result of his reflections which he intended to lay down in an extended article: "The universe as well as everything in it, is inspired with a Something, penetrated and spiritual. This Something we call God. Material is the visible, obvious expression of this Something. Everything is more or less a crystallized thought, an obvious idea, has thus a representative meaning.

Out of this Something, creatures, beings of higher order have been formed, each endowed with self-reliant awareness. In awareness the world is reflected, subjective ...

God, freedom and immortality are the postulates of the pure sense.

The fundamental idea of the plan of the world is development; discovered by Kant in the range of celestial bodies, by Darwin in the range of life forms on earth. Matter is still waiting for its saviour (It is no accident that the discovery of the most organized creatures proceeded from the fixed stars and the planets ...). The development is not a necessary, mechanical process, but instead is based on the freedom of the will ...

Development is equivalent to the expansion of the sphere of consciousness or action."

Karl von Terzaghi added to the above mentioned reflections some others, namely "human being", "intellect and mind", "human being and nature", and "mathematics". His confused ideas on philosophical problems culminated in his reflections on mathematics, where he completely mixed up mathematics and mechanics. Worse yet he planned to publish his confused ideas. On September 6, he noted in his diary: "I have happily finished my treatise 'On the Intellect'. It is the first time that I have taken up the pen. That should be the beginning and the introduction to a series of larger and smaller papers which I will attack soon." With similar words Karl von Terzaghi introduced a book, published in the middle of the 1930s, which was to be the impetus of the big scandal in Vienna.

In the middle of September, 1902, he stated that now, after all his deep meditations, he was invulnerable to all things. "I feel the power of a lion in me, since I have broken the heavy ban which encircled me for years. I know now only one goal: extreme particular education in natural sciences, a body like steel and iron and then to the farthest south."

The last days of his stay at Grundlsee were filled with hiking and climbing in his beloved Alps, sometimes with his mother.

Back in Graz his doubt about the right form of his living plan appeared again: "I have heavily sinned by my failed efforts, by my nearly outrageous meditations, although not responsible, and I am punished severely by disorder and unsteadiness. I will regain all this by the greatest strictness against myself and systematic working." Moreover, he remarked in October: "I must learn to give talks, the skill to have an effect on other persons by means of language in order to convince them with that, which I have recognized as the truth. Truth? No, I have to convince them from that, which I have inspected as right and desirable. I stand here, isolated, and will represent my opinion as the present right one, will feel myself as the center and not as a follower. My work will be to a great extent independent ..."

On October 14, he referred for the first time to his examinations at the Technische Hochschule Graz. For a long time it has been common at technical universities to split the final examinations into two parts; the first part is the so called Vordiplom examination, at the earliest after the fourth semester, and the second part (Diplom) has to be passed, at the earliest, after the eighth semester. At the Technische Hochschule Graz these examinations were called first and second Staatsprüfung. Karl prepared in a special way for his examinations. During the semester he did little or nothing to learn all that stuff taught by the professors. Only some weeks before the examination day did he start the preparation. Sometimes he locked his door for some days in order to concentrate on his learning. Finally he stated: "I am at the preliminary goal of my wishes; Staatsprüfung behind me; now I can attend to my work and my impetus. I will visit lectures. At the university I have enrolled for some natural historical studies, will, however, concentrate my main weight on geographic, geodesic and morphological fields."

The struggle between discipline and chaos in his life lasted also the next month. "In me it pushes and storms and presses unclearly in confusion. I know nothing, I think a lot, but not to the end. The unity has vanished. However, I feel a gradual mighty crystallization of my mind, a subjective feeling of self which is pressing continually more and more into the foreground."

On New Year's Eve 1902/03 he drew some conclusions of the last year. He remarked that the considerations written in his diary during the last New Years's Eve, appeared to him as an irony. "Too many intentions, too little energy. Great phrases, small thoughts. Innumerable books, lack of concentration. The year which I end today, is as each of the proceeding years, distracted. I spent a part of my time with wandering about in the dark instead of with systematic work ... However, I must admit that I made quite an imposing piece of progress this year. I have founded my philosophy of life recently through the realization of the moral law in us. I have won by this a measure of regulation and opinion of my way of acting."

At the beginning of 1903, Karl von Terzaghi considered the possibility of resigning from his membership in the corps. He did not like the present conditions there and criticized that there was no relation anymore between him and some corps brothers who were going their own ways. In an arrogant tone he stated: "This I cannot approve." In this place and also in others one recognized that Karl was a very egocentric young man. He did in no way consider the possibility that he was on the wrong road.

During March, 1903, and the next months, he was again concerned with his sexuality; he visited the brothel several times and he had an affair with a married woman. In this connection he developed some theories about the moral of sexual intercourse. But he always had a guilty conscience and sometimes despaired: "In my life I have never been in love. However, I have gone through all kinds of lust. It does in no way gratify the sexual desires, but always goes further, from intensification to intensification.

So I will let off and will seek my satisfaction in other things which do not take power but provide power. In some months, when due to the past, the view has cleared, I will write down my experiences with the whore-rabble. Lust is a broad road leading downhill. It is time that I leave it." Two days later he expanded his writings: "Lust and death are two closely related things. For this fact one finds in the world's history and in the history of an individual numerous proofs."

His dissatisfaction with the fraternity did not last very long. On June 16, 1903, Karl reported that he had fought a students' duel with a member of the fraternity on this day. On this occasion he was hurt and had to get three stiches.

At the end of July and in August, 1903, Karl was again on vacation in Kindberg, where he did a lot of hiking. His descriptions of some geological problems were very impressive. "The excursion today to Turnau through our hills has brought me some surprises. At first I found on the level of the granite ridge, along which the marked path is stretching, on some places distinct gneiss, overloaded with large glimmerplates, comes to light in almost horizontal stratification. The different kinds of rock seem to be mixed up totally abruptly and formally irregularly. The very old folds and faults seem surely not to be of a simple nature, as I thought at first. Behind *Großenbrunn* I crossed the chalk-border-line which became clearly noticeable by numerous naked stratified compacts; however, already about 600 – 700 m uphill the chalk changed apparently its look. It was in this place dark, fine slaty, extremely hard and contained black glimmer. And when I investigated in front of *Töllmarbauern*[11] some cliffs, rising vertically from the slope, I found that I had in front of me not chalk anymore but glimmer-containing quartz slate whose layers came nearly vertically to light severely dislocated and here and there nearly unrecognizably deformed ...

In the mighty, hilly basin, expanding in the western part of Turnau, one of the old, typical sea basins is present, often met in this region. The whole northern basin is filled with 50 m mighty deposits of river-brash, ..."

Back in Graz he expressed some reflections about his near future. He stated that he would be finished with his studies at the Technische Hochschule next year, and that he had achieved nothing important. "I know a little about everything and I have come to the conclusion that without specialization no advancement is possible." He considered two possibilities: The first was to visit a mining Hochschule, the second one to study geology and geophysics, his beloved field. On September 15, 1903, he wrote: "The die is cast. I am sitting at my old writing desk with the smoking pipe in my mouth and write down my decision. I will not spend my life with hunting money but I will educate myself to be a nature researcher. The future will show whether I did

[11] Peasants in the area of the Töllmarkogel.

the right thing or whether I turned down the wrong street. This choice is binding. If that what I have decided today does not become reality, I should tear and burn up my diaries. What they have contained was an illusion."

In the second half of September, 1903, he became a victim of his excessive sexual life. "Lately I was reminded, in a terrible manner, to abandon the sexual excess and I vow it personally also today. My existence, my future, the happiness and honor of the family hang on a thread." What had happened? In the middle of September he had visited the brothel and he infected himself with a chancre (veneral disease). He immediately visited the doctor who healed him with a painful method, well-known in those days. He vowed to change his life completely. However, his promise was similar to the usual resolutions on New Years Eve which serve to calm the mind but never would be kept.

In those days he was addicted to a habit which accompanied him his whole life, namely smoking. This meant a lot to him, so that he wrote in his diary on his twentieth birthday about his habit. "It (the pipe) should be my true companion during days of hard work. Its smoke banishes sorrows, doubt and inconsistency."

In October, 1903, he was again very occupied with philosophical questions. "I can say that with the acquaintance with Kant's philosophy a new area has begun for me. A glance browsing through some sections of his *Praktische Vernunft* has given a completely unexpected feeling of rest and security to me. I don't want to give up before I have absorbed everything which is offered to me there. With that a difficult, not to be underestimated task arises, namely to go exactly through everything which observation presents to me or others, and to free myself of all prejudices. An impartial view is the absolute precondition. Only a refraining from everything which previously was a silent assumption of my fantasies can level my way to understanding." Moreover, he discovered again his sympathy for student life. He praised the sight of students gathered in colorful uniforms of their fraternity with all the insignias and the student-codex. "It is a very noble thought which our student life is based on; to consider honor, i.e. morality as our greatest good and every one who wants to disown us of it we will face with the bright weapon to show him that all pain and danger are scornfully small against a dark spot on it (honor)." On October 23 he mentioned in his diary some thoughts concerning his future profession. "Now I have determined plans for the future. I will abandon all dreams of my youth and choose that profession in which I can work most fruitfully. I would like to graduate from the Technische Hochschule as well as possible in order to enlarge as ever possible, the chance to get a professorship for mechanics." This wish was fulfilled in some respects.

At the end of November, 1903, he had a passing affair with a twenty-five year old maid-servant. She had a nice and strong figure, blond hair, blue eyes and a very pretty profile.

The next days passed without incident. However, on December 17, Karl quarreled with his sister Ella. He wrote: "Yesterday my sister accused me, in the presence of a small company, of flightiness and inconsistency of character and proved this with many examples. Her words hurt me, in particular, because they contained a lot of truth. This has been one of my main sins because of which I had done very hard penance. I have, however, recognized this weakness and, to a large extent, overcome it. Possibly the German tends only therefore towards rules because his character is by nature irregular; he struggles for correction, for completion; however, he should seek such completion in himself and not outside of himself; he should purify himself from the failings of his individualism while he raises this individualism to principle."

As usual Karl von Terzaghi spent the turn of the year 1903/1904 at home. He stated: "In this night I have no reason to look back with irritation and indignation on the past year. I believe, I am writing this for the first time. It has given me again calm and recovery, it was the year of my rescue. After long, ineffective erring I have again found the right way, the human way. I have become an enemy of pure reflection, but today, at this place, they are pardonable. Work of ideas, more creative than analytic, fructified with fantasies, in greatest concentration, enjoyment without any trace of reflection. Reflection only seldom, effective like the safety valve of a boiler. Better too original than too stereotypical. Not to absorb something as totality but to educate yourself in totality. My capacity of absorption, my mobility, to heighten my individualism, to advance my unconscious forces in functions – this is the wish whose realization the following year may bring. Three cheers for the force, the art, the love, the human. A happy New Year."

The diary from 1904 is a little meager. 1904 was the year Karl von Terzaghi took to prepare for his examination (second Staatsprüfung) at the Technische Hochschule. It may be that this was the reason for neglecting his diary.

At the beginning of 1904 (January 14) Karl fought his last *Schlägermensur*[12] against a member of the fraternity, the student Re. In the morning he was very excited and took a walk. He let his lunch, which he had ordered in the Technische Hochschule, nearly untouched. "The more the moment came, the more the anxiety vanished and cheerful as seldom I climb over the uncanny dark staircase which leads to our *Paukboden*[13], took my clothes off immediately and entered, dressed in short, old, bloody military-pants and a sloppy linen shirt, the arena, sat down comfortably on a material box" in order to see another student's part at first (namely of the students R. and H.). "After half an hour he (the Student R., the author) was completely bandaged. The laying-out (of the ring) was taken up. The chalk ripped on the floor, the colophon crackled and creaked, the first commands resounded and the crowd stepped back. Both thrashed wildly at each other and R. so unskillfully that

[12] Mensur is student's duel with the Schläger (rapier).
[13] Place where the duel is held.

he immediately got a considerable quart[14] on his head in the first course. After some courses H. suffered such a sizable wound on the forehead that his second, declared him disabled. One quickly removed the bandages from him and within some seconds he was covered with blood, given over to the hands of the barber-surgeon.

Soon thereon my opponent Re. came and now we were taken to work. I felt exceptionally at ease. I bandaged the joint by myself, I protected it only by a band and in this way kept the mobility. When H. was discharged as healed, we began with the laying-out and after the common formalities I faced the opponent with the bright sword foaming with rage. Ho. (a student, the author) acted as second. I was so close to Re. that I nearly stepped on his foot. We waved for a while, each one tried to bring his sword under his opponents. And just as my sword was above his the command *start* resounded ... We trashed terribly. In the third course I gave him a solid terz[15], while in the further run I took some blows on the cheek ... Finally at 4 p.m. silentium was announced for the last three rounds and soon after that the proclamation of the accomplishment of the student's duel. I was happy that I came away unscathed, whereby I had expected a disablement. I quickly changed my clothes and walked with Ho. to my room where we closed the afternoon with a glass of wine."

In September, 1904, Karl finished his study of mechanical engineering at the Technische Hochschule after four years with the second Staatsprüfung (final examination) with very good grades. "Friendship which was built and anchored on the maxim *Blood binds stronger than steel and iron*, is blown again in all winds today. With whom do I still consort today? In strange manner with only people from time. Fleischer, the loyal and enterprizing travel companion through the Hohen Tauern, can be seen occasionally in my company at a hidden, solid and calm morning pint. Dr. Bauer receives from me (from me!) from time to time good advice for present and future and remains despite his levity the most beloved, most loyal and most reliable friend (although not in the dimension of time). On Sunday afternoon I sit together with him wrapped in a cloud of ill-smelling tobacco smoke by the good loyal, competent Dr. Ippen (Lecturer for geology), whom in the past I esteemed little as a poor, just given a right to exist researcher and in which is indeed one of the most serious human beings I know." Finally, Karl von Terzaghi drew some more conclusions: "In general, I cannot deny that nevertheless a reasonable kernel is contained in the bizare efforts of the past years. A common interest runs through all that I have experienced and dreamed up to now. The preference for natural science, often clouded, smeared, fantastic but even in periods of worst morale becoming marshy, exists. Who knows whether a benevolent fate will nevertheless put me to the place where I belong in my opinion. In recent times the idea emerges in me possibly after all to go to

[14] Quart denotes a special position of the Schläger (rapier).
[15] Terz denotes a special position of the Schläger (rapier)

Vienna after absolving the *voluntary-year*[16] in order to study natural science. Dr. Ippen said that Austrian and German geologists are gladly incorporated in the Geological Survey of North America. Could I wish myself something better? In any case, the plan is not only temporarily up in the air but rather out in space."

Also on New Years Eve 1904/1905 he reflected on his achievements and on the beginning of the end of his *Sturm-und-Drang* years. "Here I am sitting alone in my room as so many New Years Eve before. No longer a student, but engineer am I called. Thus, in the eyes of my surroundings I am better off than in the past. How one strives after title and title, after reputation and name – only to be comfortably deceived in the truth. One presses oneself in an animated way, step by step in one's progress, consciously or unconsciously, to impress one's peers in order not to have to admit to oneself how slowly one progresses and how often one takes a step backwards. The only strict good justice stays in our hearts. We have great fear in his presence and we make ourselves preferably deaf against his judgement. Every year brings a small number of new findings, in these lies the kernel, the dream of the future. The only wish which I can have is that it falls on fruitful soil. Why do I transfer the center of my world from one place to another, although its place is inevitably fixed. I am surrounded by a lot of immortal products of spirit, by an immense number of ideas and judgements about them! Am I then like the miser who collects and collects and is as poor at the end as at the beginning? For me an object should exist only then when it is able to awake in me everything that can be awaked; only then has it come to life. If I do not make use of it in this manner, it can stay or not stay, for me it is of no account. And the smaller the number and more careful the choice the better it is. We can only do that which the intellect prescribes in particularly clear moments; the remainder runs beneath the threshold of consciousness, it is not in our power. Obedience to one's own intellect and readiness to be of help to our surroundings."

[16] Ironicly the so-called voluntary-year (Freiwilligenjahr) was a compulsory one-year military service.

1.3 The Road to Practice and Adventure

With the end of Karl von Terzaghi's studies a different life began for him. He worked for some months as an unpaid trainee in a machine factory in Andritz (Graz). He recognized very soon that he would not find satisfaction in the profession of mechanical engineering. This is not surprising as his real love lay in the observation within the field of geology and the discovery of geological and mechanical problems.

So service in the army (voluntary year) in an infantry regiment was not inconvenient as it gave him time to think over the right profession for him; he finally came to the right decision. Karl von Terzaghi reported for service. He left Graz with the railway; the volunteers had to enter a simple wagon, where Karl met some friends. They discussed a lot of interesting points during their long trip. Karl took pleasure in observing the landscape. When they arrived at Steinbrück it was already rather dark. Finally, the volunteers reached their destination Reifnitz.

Military service was rather boring. Thus, he found enough time to observe the grandiose Austrian landscape during the maneuvers and he described the topography and geological problems of the landscape very comprehensively; sometimes one has the impression that Karl von Terzaghi wanted to describe the whole landscape of Austria.

He also had time to reflect on the disorder of his emotional and intellectual life. In some notes he again discussed his relationship to women. He stated finally that he always had an oppressive and shy feeling when he met a woman (Frauenzimmer). This he recognized, was based on his view of a woman as a sexual object. Moreover, he also made his disordered life responsible for the fact that he did not visit Professor Wittenbauer, his mechanics professor in Graz. "Why did I not visit Wittenbauer? Because I was afraid that this strong and serious man would put my mad confusion under the microscope."

During long days and nights on watch and the idle time during maneuvers he developed a very positive idea, namely to translate the *Outlines of field geology* by the managing director of the Geological Survey of Great-Britain and Ireland, Archibald Geikie (1906). On May 27, 1905, he layed down the concept in the draft of a letter to Geikie. "Dear Sir, With the object of filling up a strongly-felt gap in the German geological literature, I take the liberty of applying at your kindness to confer to me the right of publishing a translation of your book 'Outlines of field geology'. As a matter of fact we have no guide in German assisting the beginner in geological fieldwork in such a clear and practical way, through your kindness you could oblige many young geologist. Looking forward to a kind answer, I am yours ..."

He got the permission to translate the book by Geikie which happened in relatively short time although he stated in the first draft of a letter to a friend from June 10, 1905, that he had less time due to his duties in the regiment. In 1906, however, the German translation was published. In the

Preface Karl von Terzaghi remarked: "The natural scientific achievements are products of progress and one is satisfied to make the youth familiar with these products without strengthening the vital nerves of this progress, the capability of individual observation, and clear judgement. The focal point of the natural scientific classes is in many cases on memorizing dull factual material; valuable opportunities to bring understanding to the student are missed. Thus, one obtains a mere interest in books, systems and hypotheses instead of a comprehending love for living nature and, instead of an unprejudiced view, half an insight, in particular, blind belief in authority. One educates the youth to be receptive but not to be productive human beings. This method takes revenge in practical life later on. Therefore books, which show the way to promotion of an organic intellectual growth through natural science are necessary and deserve attention.

The university professors V. Hilbert and R. Hörnes, and in particular Privatdozent Dr. J.A. Ippen promoted this small work in the kindest manner and I thank them very much."

Karl von Terzaghi repeated his standpoint about receptive and productive learning and working again and again later on, as we will see.

On his 22nd birthday he reported that his grandfather was worried about his profession. His grandfather had heard that a boy who studied together with Karl already had the position of a government official. "And I am hanging around." His grandfather told him: "I know many people who have studied a lot and who did not make it." Karl knew his grandfather very well; he should not react. He recognized that he should show the white flag now in order to get his own way, namely that his grandfather should support his next plans. In October, 1905, he came to the insight: "I believe today is the first day in my life where I recognize the importance of my grandfather for my development, while I have considered him until now as a terrible, invincible barrier and have often regretted why I could not make my way alone. If a cool, average-born Philistine would read my diaries of the last four years he would consider me at least as mad. The recognition of the importance of the innards has gained more and more ground. Deep in us a voice is acting saying what you should do. This voice is audible in every moment of life and when we follow it always as often as it speaks, so our life is satisfied and rich in renunciation but also rich in victories. If we carry out a decision, say in scientific nature, so we have won a victory ... All that we admire in human society is a product of liberty, an obedience to an inner selfless impetus. At the beginning we recognize only the visible ... victories of this liberty, ... Today hauls us that way, tomorrow this way ..." In the following paragraphs Karl philosophized further on different points. Finally, he drew some conclusions concerning his life: "In all those things called vice I have surpassed my friends. Hardened fellows have predicted delirium for me. And almost for two years I have been nearly totally abstinent. An aversion to women (Weibern) I have never known and conscience only very little. Regret I have never felt.

And for all that there have been moments during all times where I faced all the junk totally indifferently, in which I was seemingly trapped. And in the midst of pleasure I felt often darkly that I do not really belong there and that I can leave also without a trace of regret. The results of all of these passions are experience, some physical defects, but little moral and quite no besetting vices. I have felt the question that a nature like me must wade in the morass. How could it happen. Why must I just try all the bitter foods (and they are bitter despite their enticing aroma) while L. (a friend, the author) goes without guilt and spotlessly through life? Inborn properties and education are responsible for this. On all pictures, on which I am represented as a child, I laugh amusingly, a characteristic which was surely recognized with pacification by the parents. And this foul laughing must be cured by fate ..."

On October, 9, 1905, he made the final decision concerning the choice of his profession: "I have the choice between geology and bridge building. With the money which my grandfather had earned hard and through his goodness I am enabled to learn something orderly in view of my respective wishes. I will not make further use of his support. I am intensively occupied in both professions; I must turn to them a great part of my working power. The intellect is functioning in both cases ... Geology does not exceed only description of nature; much ... work with little success. Inner relation to nature, effect weakened by the observation ..."

In the following paragraphs in his diary, he expressed some ideas about the fight of the human being against nature, and the fight between thirst for pleasure and reason.

During the winter and summer semesters 1905/1906 Karl von Terzaghi continued his studies in geology, elements of bridge engineering and railway construction at the university and the Technische Hochschule of Graz, which was financed by his grandfather for one year. He hoped that he could turn to polar exploration later which seemed to be in accordance with his inclination. However, during the summer of 1906 he had a bad accident in the Alps and it was clear to him that he would never be able to take part in such difficult tours in snow and ice. Therefore, he decided to get a job in a construction firm where he would have at least the possibility to work at times in his favorite field, in the technical geological area and his self-confidence grew steadily. The day before his 23rd birthday he stated in imitation of Descartes' *Cogito ergo sum*: "... on which I am finally able to see into the world and to think freely, completely freely, only determined by myself. I am me." And his strong *ego* remained an essential part of his character his whole lifetime.

On July 1, 1906, he was engaged by the concrete construction firm Adolf Baron Pittel in Vienna as an unpaid trainee and in the fall of the same year he reached the position of project engineer. As a mechanical engineer it was not easy for him to work on civil engineering problems at the beginning of his co-operation with the firm. He was often bothered by a feeling of lacking talent. However, in the firm there was a group of co-operative and experienced engineers who helped him to close the large gaps in his technical knowledge.

One year after his entrance into the Viennese firm, he was sent to Romania to act as an independent building supervisor in constructing gypsum silos, despite the fact that he had never seen a large building site. There he learned the elements of practical civil engineering in a very short time through a process of trial and error. "For the money, which the fast growth of my experience had cost the entrepreneur, he could have hired an experienced first-ranked building supervisor." After finishing the construction of the silos in Romania, he was entrusted with the construction of a hotel at Semmering[17], along with the conception and construction of a small but interesting hydroelectric power development in Lower Austria. In the meantime his input was appreciated and his influence within the firm grew; for example, he asked Professor Forchheimer, who he knew from his studies in Graz, to prepare an expert opinion on a building problem. Doing people favors which cost nothing, can be of great value sometimes in real life. Such was here the case as we will see later.

However, Karl von Terzaghi was not satisfied in his profession because he was not making any progress in the field of technical geology. Therefore, after nearly three years of affiliation with the construction firm, he applied for a position as an officer in the Dutch colonial troops. Before the negotiations came to an end, he was put in charge of the exploration of the preliminary hydrographic and geological studies for a proposed large hydroelectric power development on the Gacka river in Croatia, in the hinterland of the Adriatic coast. He accepted the offer and set up his headquarters in Fiume (Rijeka) on the coast and in the village of Otocac in the hinterland.

Half a year later the license to build the hydroelectric power plant went over to the *Adriatique Electricité* a formation of the *Entreprise général des constructions* in Paris. The firm gave him the opportunity to collect a lot of data for the solution of a technical-geological task. Karl von Terzaghi began to be attached to the landscape and he developed a very personal relationship to the wild and picturesque region. He hiked through the area and mapped out the most interesting parts topographically. He had learned as a building manager on different building sites in the last years to organize, to concentrate on his work and to motivate the workers. "The free life in the marvellous and wild nature and the enthusiastic devotion to our common goal made the work also in the gale-lashed wintry karst[18] into a real experience." However, he was a little depressed due to the fact that his task was limited only to a small part of technical geology; his wish for a many-sided experience became greater and greater, an experience which should include all fields of underground engineering.

[17] Mountain and resort area between Vienna and Graz.

[18] Totality of forms of permeable, water-soluble rock which are washed out by surface- and groundwater; it creates dolines (funnel-shaped holes) and caves.

Karl von Terzaghi (1907) at age 23

The results of his investigations in the Croatian karst led him to a new theory concerning the development of sink holes of the karst regions. He published his results and theory under the title: *Contributions to the hydrography and morphology of the Croatian karst* in the essays of the Hungarian geological Reichsinstitut in 1913. However, his theory was sharply rejected. First in 1918 his research results found general acknowledgement.

The cooperation with some leading staff members of the French firm was not always pleasant. "Thus, this cannot work. We cannot take up the work if always three of these nasty French men are talking." Moreover: "We must free ourselves completely from the holdup men" and "The French men have many bad habits." Nevertheless, with the aid of two French engineers he came along well with his work. However, his restless and unsteady nature led him to the decision to quit his job, owing largely to the reading of a book on hydraulic engineering, where many descriptions of constructions in North America were contained, as he wrote in the first draft of a letter to Professor Wittenbauer. In particular, there was a description of a new hydraulic method of rinsing for the manufacture of high dams. Karl von Terzaghi decided to go to North America immediately. On Thursday June 30, 1910, in the night, he rode to Graz and Vienna in order to inform his grandfather and to get some letters of recommendation for his planned stay in America. He arrived in Graz in the morning. His grandfather had his breakfast as usual in the old German

Karl von Terzaghi at age 23

room where he smoked his morning cigar. Karl explained his plans as he had done so often before. His grandfather agreed to his decision immediately – against Karl's expectations – and assured him of his financial support.

Karl spent some days in Graz, visited two friends and Professor Forchheimer[19] (Vienna), before riding back to Fiume in order to finish his work. On July 6, 1910, he sent his notice of resignation to the general management; he was happy to be ending his Croatian "adventure" in the isolated karst,

[19] Philipp Forchheimer, born in Vienna as son of a factory owner, on August 7, 1852, studied at the Technische Hochschule in Vienna and Zurich, where he passed the engineer examination (Diploma) in 1873. During the following years he worked in the fields of transportation, building management and railway construction. In 1882, he became assistant at the chair for hydraulic engineering and building construction and later professor at the Technische Hochschule in Aachen (Germany). He did his doctor thesis at the university of Tubingen and in 1889 he worked for a short time at the engineering school in Constantinople. He returned to Aachen in 1892, where he received the position of a full professor. He stayed only for two years in Aachen and then took over a professorship at the Technische Hochschule in Graz. In 1914 he followed an invitation to reorganize the Ingenieurhochschule in Constantinople. He died in Dürnstein, Lower Austria, on October 2, 1933. At the beginning of his academic career, Forchheimer was involved with problems in soil mechanics. Later he turned to hydraulic problems. His main work is the monography (5 volumes) *Grundriss der Hydraulik* (Compendium of hydraulics).

soon. Although a manager of the firm made great efforts to persuade him to stay in Croatia, Karl von Terzaghi stood firm in his decision. However, he recognized that he would miss Croatia, the karst, the Velebit mountain range and Fiume. "Farewell my Velebit. I will never forget you." He took a look at the blue Adriatic Sea for the last time and headed with the Tauernrailroad to Austria through showers and fog, passing torrential rivers and the wet Hohen Tauern, to Kitzbühel, because his family was on vacation there. Karl von Terzaghi enjoyed, after one and a half years absence, being with his family, his respected grandfather, his beloved mother and his charming sister Ella, and he wrote in his diary in September: "Grandfather was in a good mood after dinner today. I was sitting in a black armchair at the window; he paced to and for in his robe, his night-cap with the broad visor pushed deeply into his face and trained his convalescent arm in smooth, quiet motions." Grandfather told vividly and extensively some stories about his stock holdings and how he lost many shares of a rice-peel-factory in Triest due to some manipulation on the part of the management. "When I return to Graz I will sell the shares; these are the last industry shares I have. All the others are ensured government stocks, ..."

From Kitzbühel, Karl von Terzaghi started with Ella for some excursions: to Kufstein; to Munich; to Garmisch-Patenkirchen; to Seefeld; and to Insbruck, as well as Unterberg-Telfes-Stubaitalbahn and back to Kitzbühel.

In Munich Ella wanted to visit the play house Serenissimus and the cabaret Serenissimus at the Siegestor. In the Schauspielhaus with its plain, grey ceiling, furnished with only a few ornaments, Roda Roda's funny stageplay *Der Feldherrenhügel* was performed. "Roda is very witty, one could laugh. There is, however, no profit". In the Serenissimus Karl von Terzaghi was very disappointed by the program; it was too simple minded. On the last evening in Munich Karl led Ella to the exhibition park and into the main restaurant. From the terrace he enjoyed the view of the busy life and he was glad at Ella's pleasure. "She is sometimes really charming in her naivety; she forgets completely the stiffness, characteristic in our family, when she devotes herself to the impression of the moment without any mental reservation, without any convention, without any affectation. Her eyes turn brightly from one thing to another and her remarks reveal vivid participation. Yes, one can talk to such a girl in such situations. To most girls one can lie. – The innate falsehood of woman is one of the most important elements in the interplay of the sexes. The feign woman battles the man to explore how far her sexuality truly goes. It makes the man happy and ruins him."

After his return to Kitzbühel, Karl von Terzaghi did a lot of hiking. However, he found also leisure time to read some books, for example a book about technology by Max von Eyth, who lived in the nineteenth century.

On September 11, 1910, Karl von Terzaghi left Kitzbühel and headed for Brussels in order to visit the World Fair. A strange mood lay over the farewell in the dining room. His grandfather, in his brown robe and his house cap, cried. He said with melancholy: "Who knows, whether we meet again in our life – there is, however, hope." Karl could not take the deep emotion and the pain of old grandfather and went away quickly. His mother and Ella accompanied him to the railroad station. He answered shortly and brusquely when they attempted to talk to him. He entered his compartment and the train departed immediately. His mother was pale and looked sorrowfully after him. Ella was calm as always. It was a foggy day and he was in a very melancholy mood. "The train is rushing over the Inn-bridge, beneath is the broad green stream, and forward into the Bavarian highland. In Kitzbühel, however, the old man is sitting in his robe in the dining room and thinking. A long turbulent life lies behind him. He has struggled, hoped, grappled and succeeded. Then he wanted to pass on the weapons. Disappointment after disappointment he has gone through. Cheated by the son. Oskar and Victor turned out badly. Mother ill-humoured, Ella still single. Only his grandson has been developed in that direction as he has hoped, as man and as engineer. He has wasted his whole love on him. He has taught him gymnastic exercises and calculation, he has walked with him and has told him stories of Hungarian robbers and Bohemian castles. He has attentively followed his development with joy and sorrow, from a pupil and ... to a student in his final semester. Then practice has come which has quickly matured him to a man. When he came home after half a year of working in Hungary or Lower Austria tanned and hungry, there was sunshine in the house. Then he was glad about the fruits of his education and about the continuation of his life work."

The year 1911 marked a very important and decisive time in the life of Karl von Terzaghi's family and, in particular, in his own life. In this year Karl's sister Ella married Fritz Byloff [20] and Karl met on this occasion Ella's sister-in-law and his great love Olga Byloff. Moreover, Karl von Terzaghi finished his dissertation thesis and took over a new job as a construction manager in St. Petersburg, Russia.

Ella's marriage was on March 14, as he reported in his diary on March 27, 1911. Already some days before the whole house, usually very quiet, was trembling in expectation of the things which would come. Every day Ella showed, with bright eyes, always a new wedding present. Karl was fascinated by Olga Amalie Katharina Byloff, born in Marburg on the Drau (Maribor), on September 27, 1883, and descendant of an old academically educated family.

[20] Friedrich Ottokar Byloff, always known as Fritz, was born in Marburg on the Drau (Maribor), on August 8, 1875. He visited a Slovenian elementary school – although his mother tongue was German – in order to learn the Slovenian language. Then, he visited the Gymnasium in Cilli. After he had studied law, he did his doctorate thesis and his rigorosum at the Karl-Franzens-University in Graz on December, 1897. He worked as a lawyer and did his habilitation thesis in Austrian criminal law and criminal trial law (Privatdozent) in November, 1902. On May 22, 1910, he was appointed for a position as an unpaid *außerordentlicher Professor* at the faculty of law and political sciences of the University of Graz. In 1915 Fritz Byloff served as a volunteer in Worl War I and he became a full professor very late, namely, first in February 1, 1940. In May, 1940, he died.

Members of the family Byloff, from the right:
Olga, Olga's mother, father,
brother Fritz and brother Walter

Her parents Friederius Emanuell Felix Byloff and Katarina Johanna Nep. Dietz were born and died in Graz. "Olga had come from Transilvania and I saw her for the first time when she sat together with her mother in Ella's room on the sofa which was covered with a polar bear hide. A light, bright veil heightened the charm of her fine-cut fair face and her soft, clear and cordial laugh conquered quickly my aversion to the name Byloff, although her extreme esteem of the ... aristocratic circles, in which she lives, hurt me and cooled me off a little. Her slender, flexible figure stands not in the least contrast to the long ... (?) of her mother."

Gradually he fell in love with Olga. On April 21, 1911, he expressed his feelings: "Easter Sunday, Monday, Tuesday, Wednesday, yesterday, days, I cannot describe, days of wonderful shining beauty, stormy emotions, recognizing love, hours of quiet pain." On Thursday, April 20, Olga left Graz and Karl went through days of desire and agony. During this time he always remembered the beautiful days he had with Olga – and he wrote letters to Olga.

In March, 1911, he worked restlessly on his dissertation thesis and had not been able to put his daily life in order for three weeks. Newspapers lay packed on desk and closet, the pictures of his art calendar remained untouched and factual articles lay unread on the stool. The chapter on "limit strains" had absorbed his whole power. Only Ella's marriage tore him out of the productive restlessness for some days.

On May 23, he sent in his dissertation thesis *Beitrag zur statischen Untersuchung zylindrischer Wasserbehälter* (Contribution to the statical investigation of cylindrical tanks). Some professors who had to reviewed his dissertation told him that they could not write the expert opinion before October or November. Thus, his hope to be conferred the doctor degree in July and also his hope to go to the United States of America for visiting the huge hydraulic constructions in the West already in 1911 had gone.

In order to fill the next months with useful activities until his viva voce (rigorosum) he tried to get a position in a construction firm. He wrote to a friend in Riga who arranged the position in the Russian firm of contractors, J. J. Lorentzen & Sons, St. Petersburg. During the next days, Karl von Terzaghi entered into a contract with the Russian firm by telegraph; his grandfather was very happy about the conclusion.

On June 9, 1911, Karl von Terzaghi was on his way to St. Petersburg. Passing the snow-white chalk-cliffs of Semmering he arrived at Vienna where he directly rushed to the Russian consulate in order to get the visa for Russia. After this his task was clearly defined, namely to fix the foundation of the big bank building Wawelberg on Nevski Prospect in downtown St. Petersburg. It was a matter of a flat-foundation on alluvial land with layers of bog and mud. The neighboring-houses had settled irregularly due to an insufficient support of the walls of excavation. Thus, the construction had already been stopped. Because there were penalties for delay in completing the building the contractor was very interested that Karl von Terzaghi find a solution and complete the building within the proposed time. Engineer Karl von Terzaghi found the solution and the work was able to start again. During the construction work he recognized already that the settlements of the building calculated by the engineers did not correspond to the real ones – a finding which led him to new investigations in soil mechanics later.

In the following weeks he was occupied with restless working. However, sometimes he found leisure time to take a bath in the Baltic sea, to spend during the weekend a day with Lorentzen in a datcha and to visit the main points of interest in St. Petersburg, Peter the Great's legendary and wonderful metropolis at the Neva, with its special flair, the baroque and classical architecture, splendid bridges and canals. Karl von Terzaghi loved the walk along the Neva and he admired the Eremitage with the Winter Palace, containing one of the most famous art collections of the world. He traveled through the surroundings of St. Petersburg, where the lovely Kronstadt impressed him very much.

At the very beginning of September, 1911, Karl von Terzaghi finished his work at Nevski Prospect. He had succeeded with the help of 1000 co-workers who had worked in day- and night-shifts.

On September 11, he rushed for Riga, " the Nuremberg of Russia." He was charged with the organization of a branch in Riga and with the installation of a building site there. On September 11, his friend, the highly talented civil engineer Dr. O. K. Fröhlich, who had already worked in Croatia under his supervision and who would accompany him on the long way of his life, arrived, coming from Berlin. During the next days they spent a lot of time together.

In the fall of 1911, engineer Karl von Terzaghi was involved in building a castle for the uncle of the czar in marvellous Zarskoje Selo with the splendid Katharinen-Palace as well as some industrial buildings in the Baltic provinces.

The last days in St. Petersburg were filled with sight-seeing, parties and visiting the opera and theater. He described the magnificent Winter Palace and the parties extensively in his diary. At the beginning of December, 1911, Karl von Terzaghi handed over his duties to Dr. O. K. Fröhlich. On December 7, 1911, he left Russia "with a light heart" and less money. He had spent many rubles for French champagne, Swedish punch and travelling. "One month I have worked in Riga, a marvellous, old town with an easy, Russian touch. There one could live. I have wandered through Kurland, Estonia and Finland." He returned to Graz via Finland, Sweden, Denmark, Berlin and Vienna.

In his autobiographical sketches (1932) he drew some conclusions from his half-year experience: "The living standard of the workers and the efficiency of the management was unbelievably low. The corruption formed a part of the ruling system and the success in the entrepreneurship required above all a good understanding with the top people of the authorities. Any political activity was taboo. Social and professional relations were inseparably tied together and played also a decisive role in the hard competition on the St. Petersburg market raging between the German enterprise Ways and Freitag, the Danish enterprise Christiani and Nielsen and my own firm. These conditions had led to heavy economic defects, to idle motion and large loss of time. The necessity, coming from the same conditions, had, however, a motivating effect in getting to know the mentality of the clients and representatives of the authorities thoroughly and in connecting the factual considerations consistently with the psychological ones. Due to this necessity the professional work in Russia was mingled with so many purely human relations and took shape, satisfying to such an extent that I shifted my departure from the Russian work area I had grown to love from one month to the other. Finally, however, the power preponderated to expend my technical-geological knowledge by a personal approach with the practice of the American foundation work."

After his return to Graz, Karl von Terzaghi immediately met Olga Byloff again and he assured her of his love. Some days before his oral doctor examination Olga reported to him obviously that she was pregnant with his child. "During the next days that followed, I was so unable to think and to make

Olga Byloff

decisions that I did not dare to comment on Olga's information. With the sad rest of my energy I prepared for my viva voce (rigorosum) and Tuesday, January 9, the examination took place. The professors had the obvious aim to let me pass the examination. Since the function of my brain was logically and closely paralyzed to the train of thoughts of my own work, I had surely passed the vivid voce (rigorosum), however, the shameful feeling remained after all that I have given a heavy weak spot in front of the professors."

The members of the examination committee were the reviewers Professor Cecerle and Professor Forchheimer as well as the Rector Professor Wittenbauer. The reviewers had stated in a relatively short review: "The thesis discloses that he has complete command of problems belonging to the elasticity theory and that he is an expert in the field of reinforced concrete.

On the basis of the submitted thesis Mr. Ing. Karl von Terzaghi can be admitted to the Rigorosum. A working over must occur before publication because the submitted thesis is not clearly and understandably formulated and therefore it is not permitted to publish it as a doctorate dissertation." Whether Karl von Terzaghi worked over his thesis is not known today; the dissertation is not to be found in libraries in Austria or Germany.

However, a part of his dissertation was used in a book for the calculation of tanks with new analytical and graphic methods. This book had been written together with Professor T. Pöschl from the Deutschen Technischen Hochschule in Prague in 1913.

On January 16, 1912, the official celebration of his promotion took place. Many students were there and all his personal friends as well as some teachers from Güns. The natural highlight was Professor Wittenbauer's speech. "He is especially glad that he could promote, especially, his loving, devoted, loyal friend as the first one under his tenure as Rector. He has pursued my career from my first steps in practice with vivid interest." Professor Wittenbauer mentioned Karl's activities in Vienna, Transilvania, Lower Austria, Croatia and, in particular, in St. Petersburg. "And what makes me, in particular, happy is that he cultivates in such a careful manner our German mother tongue despite the great demands incessantly raised by his profession. The result is the beautifully, clearly worked out style so that we could hope he becomes, in days to come, like our adored past master engineer Max Eyth. And he closed with good wishes for the further career of the young German cultural pioneer."

Together with the senior of the Vandalen, Dr. techn. Karl von Terzaghi entered the carriage and rode home. There, Ella, Fritz, mother, and grandfather were waiting for him and grandfather presented him a splendid golden watch in memory of the day of his promotion with trembling hand and with suppressed emotion. "I let quite a row of watches be delivered from Schaffhausen and I chose the best one. It has cost 300 Kronen. You will be able to put it to good use."

After his promotion he had a strange and mysterious affair with the wife of an acquaintance, beside his passionate relationship with Olga Byloff. However, he brought the affair to an end within a few days and said farewell to her. "I will never forget you in my life", he wrote

Due to grandfathers's cold he had to postpone his departure to the USA and stay in Graz until Monday, February 12, 1912. "On Sunday evening I sat together with grandfather in the café Vienna. I had really the wish to spend still some hours in his company before I leave him for years, yes probably forever. I felt obliged to have none of my wishes undiscussed and none of his intentions. Should have a clean slate. Thus, I told him that it is my intention to bring Olga to New York and further all that I knew about the sad point in Ella's matrimony. For just this and Ella's helpless pain had heavily shaken me during the last days of my stay in Graz ... Grandfather was obviously touched painfully by my information, however, he is too wise to express his feelings somehow unwillingly or vehemently." At the end of their conversation his grandfather informed him about their family's pecuniary circumstances: for Karl 150,000 Kronen, for Ella 150,000 Kronen, both parts with full interest profit. "Mom has the house with 3,400 Kronen proceeds and the interest profit of the total means." Later Karl forfeited his inheritance of his part of the house in favor of Ella. In a letter to Ella he recalled the time they had lived together in Graz: "We have spent together our childhood in the house and all memories are connected with you; therefore it stands to reason that the house remains in your hands."

On Monday, at noon, Dr. techn. Karl von Terzaghi departed from Graz for his great adventure, to become acquainted with the American way of life, to widen his knowledge in underground engineering, to explore the American landscape and geological conditions, and to collect all data connected with underground engineering and geology. His grandfather gave him a kiss. "We hope to see each other again." Then he went into his room. His mother, Ella and Fritz accompanied him to the railway station. It was a sentimental farewell.

Karl rode in a sleeper section from Villach through the Tauern to Munich, then further to Hamburg. On Saturday, in the morning, in Hamburg's main station the train was prepared with wagons of I. and II. classes picking up the passengers for Cuxhaven. On Sunday, February 18, in the morning, the steamer *America* set slowly in motion. There were stops in Southampton and Cherbourg embarking new passengers. Karl von Terzaghi enjoyed "dolce far niente", due to the generosity of his grandfather, in the first class: "breakfast – music – lunch – music – dinner – music" and became acquainted on board with some interesting people. "Every morning a bath at 8:00; then into the gym, in which one can train the muscles, ... Music accompanied the small brunch, beef tea on board and to the meals a menu of generous richness, no limit for the number of meals ... Daily specialities of oyster, lobster and exotic salads, pineapple."

On February 23, 1912, on board, Karl von Terzaghi wrote a puzzling story about a trip with Olga Byloff to Budapest. After his vivid voce (rigorosum) he had time to think over his relationship with Olga. "From the first moment when I got closer to Olga, I had the unclear feeling, if Olga is still a virgin, than by a lucky pure chance. Her whole nature is orientated towards receiving and demanding sexual love. I have not given to her, due to conscientiousness and respect for her future, that which her nature has demanded; I wanted to condemn her against her nature to abstinence and the consequences are not wanting. When I withdraw myself now from her, my whole sacrifice was for nothing and I bring her into the arms of others, into the arms of men who give her that which her lover has refused and expose her to sure downfall. Due to this I have no alternative but, to rescue her by my personal influence, by my unselfish and pure love embracing the whole girl, than to make up for that which I have missed. She should become acquainted with the difference between vice and love and through complete union with me be saved from the enjoyment of distasteful surrogate[21]. This insight ripened soon to a strict decision and after repeated exchange of letters (in this course I had to conquer her first, great mistrust) we made up our minds to meet in Budapest. In her habit after the first letter, in which I offered her complete pardon and love, the characteristic trends of her nature appear clearer than ever before. She did not play the hypocrite, she simulated no passion, but wrote crystal-clearly that she has to consider it, by the way a habit that had driven me nearly to

[21] He means substitute (the author)

frenzy in that moment. However, possibly the supposition lay close, that I wanted to rush at the supposed victim like a vulture. She showed her inborn selfishness and her lack of imagination, because she shows no sense and no understanding for the consequences which the mental benevolence in the last weeks had to handle and she behaved nearly indifferently in view of my successes."

Karl von Terzaghi continued this strange story on February 26, 1912, while still on board. "Budapest. After a rather uncomfortable night in the sleeping coupé an unpleasant morning at the central railway station. The ... train was delayed two hours due to ice and cold and I nervously expected its arrival. I promenaded to and fro on the platform with very pessimistic thoughts in my mind. Finally, the train arrived in the hall and some minutes later I led Olga for breakfast into the next café. The first minutes were icy, something like mistrust laid between us. However, soon after the first discomfort was overcome the old feelings and senses came back more clearly and strongly and lovely than ever; when we left the café, I felt that the point was reached to drop the last barriers. Olga, when we stay at the hotel Royal we are man and wife, do you understand? 'As you wish,' Olga replied without looking up and without changing colour. Thus, it had happened. When we entered the elegant and spacious room in the third floor of the hotel ..., I had the feeling we had been married for years."

They spent the rest of the weekend together in Budapest visiting some places of interest. Karl reflected his close relationship to Olga. "Olga was of devoted tenderness without reserve and had no other wish and no other feeling belonging totally to me and I took her as legal property." He continued his sentimental writings with: "I thought over the manifold changes of destiny which had led me, in different life situations, always again to Budapest and I felt happy being in total possession of a girl who I have become attached to not only as a woman but also as a young friend in need of help and as a human being." The recollection of the visit to Budapest abounds in Karl von Terzaghi's selfishness and egotism, in particular, in the last paragraphs – and in a patronizing macho tone. It contains many contradictions and he did not tell the true story. The question arises: Why did Karl von Terzaghi write this story? Maybe he wanted to overcome his frustration concerning Olga's pregnancy by writing down the tale. One has to consider that the whole affair happened before World War I when the morality in Austria and other places of the world was stuck in the nineteenth century and an illegitimate pregnancy was always a scandal in those days. However, much more probably it appears that he wanted to suppress the unexpected pregnancy, this "dark point" in his life, because of his figure in history, of his stature. He liked, also during other occasions as we will see later, to gloss over unfavorable occurrences in his life. This he did with great care. It is astonishing that such a personality, such an intellectual man, really thought he could make such "dark points" in his life disappear.

After ten days he had crossed the Atlantic Ocean and he arrived in New Jersey (Hoboken) on Tuesday, February 27. "A blood-red sunset, emerald green sea, clear sky and a sharp, fresh wind from the north". Around nine o'clock he came to Manhattan and was overwhelmed by the view of the concentrated mass of buildings with the huge skyscrapers. "Light, light, light as if the whole storage of light of the earth" was concentrated in Manhattan.

He stayed for two days in the Astor hotel, one of the finest addresses in New York, because he could not find a middle-class hotel. Then he moved to a boarding house in 76th Street, to a friendly room, with breakfast and dinner for a weekly rate of ten dollars. The boarding house was located in a very quiet street close to the splendid Central Park and the American Museum of Natural History.

The first weeks were not very encouraging in the New World. One day, after an extensive tour of different bars he was very drunk. "On the next morning I awoke with a headache and felt lower than I have felt for years. Almost downhearted I walked for hours between the ponds and rock hills and pine trees of Central Park and the desolation of my New York existence came to my mind in terrible clearness. I was ashamed of my unemployment and was ashamed to have wasted hours behind the wine glass without earning money during the daytime. I felt like a useless being in this huge town of producing and acting and I decided in this state of deepest dejection to try everything in order to get a job as soon as possible. The thought of Olga strengthened my feeling of depression to intolerableness. Due to my debt and my carelessness, against the better conviction of relatives, two useless creatures in this huge town, ..."

Obviously, he had come to the USA with too great expectations and the disappointments continued. "One is not waiting for me. Quite on the contrary. European experience and, in particular, European theory are at a discount. I suffer fiasco after fiasco and I am today still just so without job and without well-grounded prospect of a profitable occupation as on the day of my arrival. Only poorer." He came up with the clearly logical notion that the New York people were responsible for his disappointments. And he thought over his failure for a long time and came to the conclusion: "The European engineer is, in general, educated too little rationally and works too little goal-orientated. This perception, considered from a financial standpoint, begins to penetrate my whole personality, to influence my way of thinking and to bring clearness into the theoretical part of my wishes ..., it is the wide, wide country in the west which guarantees the final success with space and with tasks of any kind."

In the west of the USA a large-scale irrigation program was about to commence, initiated by President Roosevelt in the *Reclamation Act* of 1902. It included the construction of sixty dams. The United States Reclamation Service was entrusted with the realization of the huge project. At the dam sites, the engineers had to deal with nearly all technical-geological difficulties which could occur in the field of civil engineering. Thus, if anywhere in the world there was an opportunity to create a broad basis for the development of technical geology, it was provided by the systematic working out of the experiences which were being collected at the building sites there.

Karl von Terzaghi was already close to fulfilling his wish. On March 22, he visited Mr. Plant, who he had already met on the *America*. Mr. Plant handed in a letter of introduction to the head of the Geological Survey in Washington DC and invited him to the wealthy Chemical's Luncheon Club in the Williams Club where they had an opulent luncheon with oysters, morels, haddock and strawberries. On Saturday, March 23, he finished his article on the graphical method of investigating the tank bottom (his dissertation thesis) and he stated that this should be his last theoretical work. With the letter of introduction, Karl von Terzaghi rushed to Washington DC. However, his first appearance was not quite encouraging. "I came into uncertainty." He expected elegant houses in a very modern town. However, he wandered about one hour through the streets of a workers' settlement. Finally, he found board and lodging for two dollars in a hotel. On Thursday, March 28, Karl von Terzaghi went to the head of the Geological Survey. "What can I do for you?" Karl explained that he would like to study dam constructions and irrigation problems. "That is Reclamation Service, Mr. Newell ... Please go to the man." Then he wrote some lines on Mr. Plant's letter of recommendation. "Mr. Newell, an old Gentleman with small bright blue eyes, round headed with a hooked nose welcomed me with the calm politeness of an expert. I briefly explained my plan, to study the bad and favorable experiences gained during the construction of the large irrigation projects. He seemed to recognize my plan as very practical, absorbed the subject with liveliness and in a short time I was informed about the most important projects of the far west; with American realism he promised me travel folders and letters of recommendation."

On March 31, still in Washington DC he worked out a report for *The Österreichische Ingenieur- und Architekten-Verein* (The Austrian Society for Engineers and Architects) in order to get a grant for his proposed trip to the irrigation projects in the far west of the USA. Karl's attempt to raise money from this society appears a little strange. "I have no hope of success, however, the attempt is worthwhile, a new experience is worth a working day." In this report he outlined the purpose and his program, which contained twelve projects, for the study trip. The purpose was to study in the far west dam and canal constructions and working methods and he listed twelve projects he wanted to visit, e.g. the Rio Grande, Yuma and Yohina projects. The costs for his trip New York – Chicago – El Paso – Vancouver – San Francisco and back to New York and for the three month period of visiting would be 1,300 dollars, which amounted to 6,500 Austrian Kronen at that time. However his proposal was not successful. *The Österreichische Ingenieur- und Architekten-Verein* rejected his proposal as Karl reported later.

As already discussed with his grandfather in Graz, Olga came to New York. Probably Karl's grandfather paid for the voyage. However, there is no evidence for this assumption. On April 2, 1912, Karl von Terzaghi met his pregnant and unhappy Olga. He went in the dark in a downpour of rain and in dense fog to the pier to welcome her. The piers of the Austrian company lay in an unkempt, obscure part of Brooklyn and two hours ride from Central Park. He accompanied her to his neighborhood; and she enjoyed boarding in the same house as Karl and Karl's landlady took care of Olga in the finest manner. Karl and Olga spent the next four weeks together in New York. They could meet and do sight-seeing or visit acquaintances only in the evening or on the weekends because Karl had taken a day job as a draftsman. Sometimes they sat in his room and Karl talked with Olga about her situation.

At the end of the four weeks, when Olga had already left New York he reflected on his emotions, his happiness, his sorrows and Olga's sudden departure in a draft of a letter to a life time friend in Graz: "And it was an idyl which I had lived with Olga for four weeks. The idyl was disturbed by nothing. Day by day I worked like a carner on the building documents of the restless, cranky Yankees, however, the evenings belonged to us. Olga remarked yesterday on the way to the Pennsylvania railway station with tears in her eyes, that it was the most beautiful time of her life".

During the last weeks Olga had tried to get a position through charity institutes. However, she failed and Karl got very nervous. Finally, Karl introduced Olga to the Austrian general consul on recommendation of a personal friend of the consul and she got a warm letter of introduction. The consul gave her the advice to ask the Sisters of Joan of Arc and she was successful. On Thursday, April 25, she was informed by phone that a position was available, and that she should come at 5 p.m. Later Olga told Karl on the phone that a rich Mexican plantation owner, Mr. Ramos, who lived near to Mexico City, wanted to hire her as a governess for his three children. His wife had been very ill for years and he expected that she would die very soon. Mr. Ramos invited Olga for dinner in the hotel Hofmannshaus, Madison Square, and Karl might also come. He freshened up and rushed for Madison Square. In the dining room Mr. Ramos welcomed Karl von Terzaghi with a strong handshake. Mr. Ramos was a stocky man of around 40 years with an interesting profile and an open, honest face, full beard and dark-brown eyes. He interrupted his meal, ordered a bottle of Rhinewine and started immediately with great liveliness to explain his intentions, wishes and his circumstances. He was obviously endeavoured to convince Karl of his prosperity, of the spotless reputation of his family and of the absolute purity of his intentions. With exquisite politeness he accompanied both, Olga and Karl, to the subway.

On the next evening Karl had a talk with him in his apartment and he demanded a contract and references from Mr. Ramos. The Mexican promised this with ardor. Moreover, Karl visited the Mexican general consul and he got the best information. The general consul also weakened the wild rumors about the Mexican revolution. He assured him that the turmoils were limited to the northern border, 1,000 kilometers away from Mexico City. Based on this information Karl von Terzaghi agreed to the employment.

Olga rode together with Mr. Ramos to Mexico. From St Louis Karl received a telegram "your dear sister arrived safely we continue trip tonight; regards Juan O. Ramos." Olga and Mr. Ramos arrived at the plantation in Mexico on Wednesday, May 1, 1912. She earned 3,000 Kronen (600 US dollars) per year. This was a respectable amount considering that lodging and boarding were free so that she could save a lot of her salary.

Karl was alone in New York and free. He could prepare for his big tour through the west of the USA. He still found, however, time and leisure to write features for the feuilleton part of Austrian newspapers as he already had done in Graz. One article was concerned with the sinking of the Titanic. "I believe, the time is not lost. My imagination is promising and training can be useful. It would be for the time that I describe an occurrence which I have not gone through and landscapes which are foreign to me." At the end of his stay he added a further article to his already written feature, namely about skyscrapers. In this article he thought about the difference between the modern and the old towns.

His financial situation was not the best. *The Östereichische Ingenieur- and Architekten-Verein* had not responded to his proposal. However, he could rely on grandfather's financial support. In a letter to his grandfather of April 28, he returned thanks: "I have received the letter of credit four days ago; I thank you for your goodness, with which you support my project and I will make efforts on my part to extract as much profit out of my trip as in my power." Moreover, to his surprise, the Royal Hungarian Geological Reichsinstitution contributed 500 Kronen to his scheduled trip to the Southwest.

On Saturday May 4, he received the first long letter from Olga wherein she described that Juan Ramos took care of her with all of his tenderness and with loving attention. Karl got very jealous and he stated angrily: "I could kick myself."

Tuesday, May 14, 1912, was Karl's last day in New York. Olga's situation had obviously dramatically changed. In a long desperate letter, which he received during dinner, Olga told Karl that she was very unhappy and complained that Mr. Ramos made advances to her. Despite cleaning and packing Karl answered Olga although it was already midnight and he criticized Olga: "You did not follow my advice. From the beginning you had to remind Mr. Ramos with emphasis on his duties as a gentleman and you should nip in bud with iron persistence each advance." At the end of his letter he assured Olga that he would come to Mexico in August or at the beginning of September and stay for two weeks close to her.

On Wednesday, May 15, he rode with the express train to Washington DC, where he again met Mr. Newell.

He left Washington DC on Friday, May 17, and headed via Atlanta for New Orleans, the Big Easy, where he arrived on Sunday morning. It was a terrible ride. In the first night, drunken passengers were very noisy, the train sometimes abruptly stopped and moved so that Karl could not sleep during the first night. The next day he saw along the railway villages with primitive wooden houses founded on isolated brick foundations, red painted, almost neglected and collapsed, abandoned farms. Karl was not pleased, at the first glance of New Orleans. "Dirty, old-fashioned, primitive railway station close to the stream, the rails unprotected between docks and town. Canalstreet, broad street, 5 – 6 tram rails one upon the other, densely occupied with neglected, orange tram wagons, 4 – 5 floor high houses, stone or brick facade, ..." He took a walk along the dirty docks; the harbor reminded him of the harbor in Fiume (Rijeka). In the evening he enjoyed a marvelous sunset and admired the heavy, square white Mississippi steamer with paddling wheeles. In the French Quarter he felt the special atmosphere of the Big Easy. "Houses low, grey, unclean, with fine, old wrought-iron balcony railing and fine pretty small columns and loggies. Like in old Venice. In the garden in front of the temple Sinai shining flowers, very old fan palms. Strange contrast."

On Tuesday, May 21, he had the opportunity to view the closing work of a breach in a Mississippi dike 40 km north of New Orleans. In March the dike had failed and water poured through a breach 330 m wide and flooded in the next weeks numerous sugar plantations. Karl von Terzaghi vividly reported about the failure and closing the breach in an extended article in the magazine *Die Umschau*, 1912.

At the end of May, Karl von Terzaghi was on his way to the South-West of the USA. He rode through Texas to the Rio Grande areas, to El Paso through bush land close to the Rio Grande. The railway dam was 0.75 m over clay water and secured against flowing water by stones. On his way he saw Mexican hovels, alfalfa, and wheat fields. The ride lasted 48 hours and was very exhausting. However, on the next day he visited the head of Angel Dam together with the department of the United States Reclamation Service (U.St.R.S.). "Excellent building management I have seen up to date. Perfectly arranged."

The first major stop in the South-West was Elephant Butte in New Mexico. On the ride to Elephant Butte he passed Lanstry and Fort Seldin where he watched a railway accident. The railway was blocked and after many hours of waiting, the passengers, who showed much patience, were brought with a luggage wagon eight miles to the north, to Rincon. There, the passengers entered the train again and in the evening they arrived at Butte Junction. In Elephant Butte, close to Butte Junction, Karl von Terzaghi visited the dam site. After lunch he took a walk with his hosts through the spare landscape with low cypresses overlooking the mixer unit. The people at the site were very busy. A locomotive engine brought excavated material on the Cofferdam. Some arc-lamps glimmered above the chasm on the opposite side of the Rio

Grande. In the evening after the walk, Karl played a game of billiards with his hosts. On the next day Karl sent some typical stones to an acquaintance, a mineralogist. In this wild solitude he took pictures of the vegetation, of the geological situation and of Elephant Butte from the Ash creek side and he recorded many details in his diary. In the evening he had dinner with one of the construction supervisors and later he worked on an article entitled *Camps in Desert and Lava.* In an entry in his diary from June 7, 1912, written in Phoenix, Arizona, he remarked that he had finished an additional article for a newspaper in Leipzig with the title *Kriegsschauplatz* (Theater of War).

On Sunday, June 2, in the evening, he arrived in Tuscon where he stayed until Wednesday, June 5. He visited the university on Monday and was impressed by the spacious and clean campus. Moreover, he attended the graduation ceremony and enjoyed the celebration very much. On Tuesday afternoon he made a tour by car[22] together with some acquaintances and their families and a professor of civil engineering in order to view a pumping station on the Santa Cruz River. The car raised much dust; the ground was fine like cement. Later Karl von Terzaghi rode eight miles to San Harin. "The dusty desert with the unfriendly clump of trees acts as a deterrent ..." Beneath the grey sky they saw an extended building with a small dome. "Some low Indian hovels to the right of the street. A priest was talking with two Indians. Two warm-hearted fat nuns opened the gate. The room with colorful Indian baskets, on the walls large photographs of important Indians in war decoration. The interior of the church, gold, colorful, poor pictures, though, good carvings most made with wood of the Mesquite tree. The carvings must have made deep impressions on the Indians when they saw them for the first time, the grieved faces in the dimness between gold and color ... The portal with simple copper fittings reveals important wood work. The middle piece of the facade over the gate looks like an old bearded sculpture ... Small garden in front of the church with iron bars and, to the side, an Indian cemetery."

At the University of Arizona, Tuscon, Karl von Terzaghi admired the valuable ore collection. "The most beautiful malachites, azurites, molybdenite etc." Moreover, he visited a testlab where experiments with materials for road construction were performed.

On the morning of Wednesday, June 5, he rode on a horse to the Desert-Experimental-Station, one-and-a-half miles southwards of Tuscon. The ride took two hours on stony paths over two hills. In the station, engineers, chemists, biologists and physiologists were working. The experiments were concerned, in particular, with the influence of solulable soil salts and evaporation.

From Tuscon, Karl von Terzaghi travelled to Phoenix on the same day. During a stop he saw some charming Indian girls and some Apaches with serious, energetic faces. During the trip he was surprised by a tornado and

[22] Karl von Terzaghi did not possess a driver's license and he did not apply for one during his lifetime; he always expected that his hosts would give him a ride.

observed yellow columns of sand of approximately ten to fifteen meters in diameter at some places.

In Phoenix Karl prepared for his adventurous trip to the north, to the Colorado area. And on June 8, he climbed down the Grand Canyon (Colorado Canyon), was impressed by the uniqueness and magnificence of this world wonder and he scarcely dared to enter the rim of the canyon in the loneliness and the "terrible monotony." From the Grand View Point he had an overwhelming view and he began to feel the "terrible solemn mood" of this canyon which the Colorado had formed over the course of millions of years. Of special interest for Karl von Terzaghi was, of course, the geological structure of the Colorado plateau. "The view through the breach in Lower Tonto of the terrific dark primary rocks is incomparable. The idea of the huge age of the basis of the Colorado plateau suggested itself. A world seems to lie between the Archäum and the Lower Tonto. The view of the isolated mesas beyond the stream with the sharp red stages and the leek green slopes recalls vividly our Dolomites." Karl investigated further the layers of rock during his trip down to the clay-banked yellow Colorado River with an Indian guide. In the Indian camp with several small wooden houses at the oasis in flourishing gardens, he met two remarkable young men. They presented him with a bottle of water. One of them spoke German. The stranger had immigrated as a three years old child from a Rhine province in Germany. After a while, Karl von Terzaghi hiked up and after an exhausting walking tour of ten hours he reached the plateau. Back in Phoenix he again set out for more sight-seeing; for example, he viewed the Roosevelt Dam and the Salt River Reservoir.

In the middle of June, 1912, he travelled to Yuma by car and by train after a strenuous ride and after having visited a dam under construction. By the curb sat many Indians. "Open dirty lunch halls, Chinese barkeeper, ... saloons more neglected than in Phoenix, suspicious, drunken faces with glittering eyes, intruding penetratingly ... Along the dike long lines of Mexican mud cottages, almost exclusively open, freshly covered shed, beds outdoors, often equipped with mosquito nets, fences of barbed wire, often neglected, dirty children."

On June 18, Karl was on his way to Los Angeles via Mexicali (Mexico) and arrived there in the evening. Over the next days, accompanied by a hydraulic engineer, he enjoyed, the marvellous surroundings of Los Angeles, in picturesque Southern California.

After his extended tour through the South-West, he travelled with the Southern Pacific railway company along the fantastic, rocky coast to San Francisco on June 25, 1912. "The Promised Land, California was traversed – however, the expected surprise, the general and solid prosperity, the guarantee of improving one's position for every healthy man who is willing, failed to appear. In Los Angeles I did not feel the defects of the country yet. The beauty of the town and the carefully exploited productivity of the land may be mistaken in this. However, in San Francisco it is hardly possible to hide."

During the next days, Karl von Terzaghi headed for the Sierra Morena and to Mount Hamilton to the south of San Francisco. His first stop was in Redwood City. From here he inspected the countryside, 18 miles westwards of his stop, where he met a farmer from the small village of La Honda. He invited Karl von Terzaghi to ride with him to Woodside and La Honda. On the next morning he went by foot back over the pass to Woodside. On July 5, he stayed in San José, where he wanted to rent a horse for his intended ride to Mount Hamilton. However, it was difficult to get one because the people told him the way was too rough; however, he succeeded. "On the next morning I rode at a sharp trot and gallop within two hours to New Almaden in the east, 14 miles." In the afternoon he left New Almaden and by 6 p. m. he had reached his hotel where he led the horse, its whole body covered with salt, to the stable. On the next morning he rushed with a rented car on endless serpentines to Mount Hamilton. "Although the mountain shows no notable parts of rocks, the slopes appear nearly threatening with the isolated dark oaks and the glens deeply torn in the flanks." Karl enjoyed the nice view of San Francisco Bay and San José very much. He climbed a plateau of the mountain. The observatory was located in the center and Karl entered the building and admired its enchanted world.

It was already dark when he climbed down. He was a little careless and suddenly he lost his footing, took a tumble and lost his pince-nez. The next morning he happily found his pince-nez and in the afternoon he was back in San José. After further excursions the time came to say good-bye to San Francisco. On July 8, 1912, he left with a steamer for San Diego. The coast land reminded him of Italy's coastal landscape. The steamer stopped in San Pedro, a small, new town in the southern Los Angeles area. On board he met an old man from Pittsburgh and a farmer from Wyoming. Both were simple but intelligent men. In San Diego, southern California's most beautiful city, he went with them into a hotel and spent the next two days in their company. They took a nice trip to Tijuana, in some places dirty and dusty. On their way back they drove along the Coronada beach with small strips of yellow sand dunes. A woman from Ohio, 200 pounds heavy, talked nonstop about the conveniences of the North-West, of Portland's roses and of the Indian legends from the Columbia River. At the end she stated: "I must die, if I can't speak."

Karl was deeply disappointed by Point Loma. This may be on account of the fact that this area was under construction in 1912. However, the view from the west of Point Loma to the harbor of colorful San Diego, with the marvellous skyline attracted him to this region. On the next morning, Karl von Terzaghi visited Coronado Beach and was impressed by the Coronado Hotel, which later became famous as backdrop for a scene of the film *Some Like it Hot*. "The hotel with high red tiled roof, with turrets and garrets does not fit perfectly but nevertheless well into the beach installations." Moreover, he compared Coronado's sandy beach with that in Los Angeles: "The same sandy beach as in Venice, Santa Monica and Long Beach, where I had seen the Pacific Ocean with its thundering, rolling surf and its subdued, fine blue for the first time in June."

After San Diego he was back in Los Angeles. "That's L.A., is California all over and must be studied. During my first short stay I had become acquainted with a row of more or less interesting human beings." Most of his acquaintances had to do with water supply and they took Karl von Terzaghi to many interesting constructions sites, related to water supply, sewage and irrigation close to Los Angeles.

On Saturday, July 13, he was on his way to San Pedro in order to go boating to the Santa Catalina Islands. "However, Santa Catalina, the island. Completely different from, e.g., Capri. The coast rises steeply out of the dark blue sea, however, the colour of the rock is dark green with a tinge of sulphureous or rust red. Dense bush covers the moderate rounded hilltops, partly straw yellow dried grassland and luxuriant, tropical vegetation in the canyons sprouts up high, in the background of the bays with brilliantly bright pebble beaches." Karl von Terzaghi mostly spent his time relaxing, e.g., bathing in the sea and doing sight-seeing (visiting the undersea gardens).

Back in Los Angeles he visited some acquaintances who had invited him to lunch and dinner in Pasadena and Venice. Moreover, Karl rode with the train, accompanied by an acquaintance, to the Coachella Valley, close to the San Bernardino National Forest, south-east of Los Angeles. On Saturday, July 20, 1912, they reached Indio. At that time Indio, as well as Coachella, the next stop, were still small villages. Then he headed for an Indian reservation which he harshly criticized as dirty. "The government takes care of the people, bores wells, gives machines and oil, however, the people do not make progress." In their very primitive lodging they had to watch out for snakes and scorpions. From Mecca they saw in the distance the yellow Colorado desert, the split slopes of the San Bernardino mountains and the Salten sea, a bright blue wonder. A wooden shack served as a hotel. Karl's goal in Mecca was to visit an experimental station for irrigation problems. In the time to follow, Karl von Terzaghi tried to get involved in two business projects, in order to participate in the growing economy of Southern California – and to follow after his grandfather's footsteps.

On the ride from New Orleans to El Paso, at the beginning of his great excursion to the West he had become acquainted with a New York banker Mr. C., a thin, flapping man with grey, blank eyes and the face of a prater. In Yuma he received a letter in which the banker called Karl's attention to the Imperial Valley, just north of Mexicali at the Mexican border. "With this valley we can make some money," the banker stated. In San Francisco, Karl von Terzaghi met him again by chance on the street and from that point on he was continuously in contact with him. His ancestors could be traced to old European families and they had served in nearly all the wars of the United States. He had made a lot of money twice. However, he also lost it twice. "I will tell you something, Doctor, about cotton. Cotton is the single business in the world that's absolutely sure. You go to a broker in Liverpool and tell him that you have cotton in California and you get cash. Cotton is money, cotton can be shipped where you like and cotton can be stored. You and I, we have the key about the Imperial Valley in our hands. I know everything about cotton, you know everything about the irrigation. I with my friends, and you with yours, we can make big money."

One day Mr. C. introduced Karl von Terzaghi to Mr. Major W., a tall man with a genial, open and energetic face. He traded in real estate and he was making great efforts to sell the land of a ranch. His plan was as follows: he wanted to parcel out the land (4,000 acres), to sell it and to cultivate it in exchange for 25 percent of the profit. On the basis of a rough estimate he calculated an enormous net profit. However, Karl recognized very soon that the whole calculation was wrong. There were no considerations for cost of upkeep, no calculation for the necessary, expensive agricultural machines, no knowledge of water rights. Mr. W. referred to Mr. E. who was supposed to have the corresponding files. Mr. C. introduced Karl von Terzaghi to Mr. E., who seemed to be a good-natured man accustomed to comfort. He stated: "Look, I studied the situation for two years before I chose 16,000 acres from the huge land complex." However, Karl von Terzaghi got the impression that the legal situation was chaotic and, finally, his efforts to participate in the business failed. "I was vividly interested in his projects and had checked the technical and legal basis. He had prepared the work with scientific thoroughness and he knows his business. The proposal which he had made was to participate in the enterprise with a share of 1/8 by paying 100,000 dollars; this is truly an attractive offer and means a fortune for the investor. I am sorry to say good-bye to the honest, serious hard-working man."

Karl von Terzaghi's second attempt to go into business started in July, 1912. An acquaintance had told Karl of his intention to plant date palms on the two properties in the Coachella Valley. Karl was vividly interested in date palm plantation. He recognized the extremely favorable location of the Coachella Valley between the Imperial valley and the San Bernardino mountains. He persued the plan to rouse the interest of Hungarian capital for a 500 acre project. The encouragement for this came from a piece of information from Mr. B. that there was a permanent spring at the north slope of the San Bernardino mountains, only 12 miles distance from the proposed land away and 45 miles away from the railroad. The costs for the water pipes would be 40,000 dollars and the total costs for the project 1.500,000 dollars. He hoped to make about the necessary connections in Hungary with the help of the Imperial Geological Institute. Simultaneously he thought out a combination as an alternative project, namely to ensure 30,000 dollars. With these 30,000 dollars he wanted to start the project on Mr. B.'s properties and to participate in the consulting of the developer as an assistant. However, this

Karl von Terzaghi, Coachella Valley, California, 1912

gentlemen's business was so meager that he could not afford an assistant. On the day that he put forward his second plan without getting the desired result, the developer welcomed Karl with a mysteriously pleasant face. A group of men had come together and had developed an excellent project. The only thing that they were missing was money. The negotiations were very interesting. However, in the end they fell through.

There was a little epilogue. He learned that an old prospector had rejected an offer of 35,00 dollars from Southern Pacific authorities for his farm. The farmland seemed excellently suited for date palms. However, the negotiations about the purchase of the farmland failed also and, finally, he closed the book on this chapter of his life: "The dawdling has come to an end now. The last month had fallen out of the program; however, it has given me a valuable insight into the way one can make money in new countries."

Gradually he got a little homesick. "A letter from grandfather; the thought of him touches my heart and it hurts me to know that he is alone. Fritz's letter: he thanks me for my efforts concerning Olga. The poor girl."

With the two attempts at getting into American business Karl von Terzaghi's stay in Los Angeles came to an end. After visiting some acquaintances he said farewell to L.A. and rushed to San Francisco arriving there on August 8, 1912. During the first days in San Francisco he reflected on his experiences in Los Angeles. In a first draft of a letter to Professor Wittenbauer he complained of the bad conditions for engineers. After some thoughts about the good earnings of speculators and workers he stated: "Only one kind of person is banned from the banquet, the engineer.

This is an interesting chapter. I have collected data and facts everywhere. Only one example: On the oil fields of California one pays for a day-labourer 70 dollars per month ..., for a foreman, 200 dollars, for a supervisor, 400 dollars. Engineers work in the first 5 – 6 years, yes even in the first 10 years, for 60 – 80 dollars and experienced men for 150 dollars ... He, who adopts the profession of engineer is an idealist. Therefore the unusual solidity of the American engineer within swindling piracy; for joy in his work is the only thing he has."

Moreover, he complained of some bad incidents during his stay in the United States. "However, I cannot go back" – and he remarked "I see my future in rough drafts in front of me. The easily gained riches of the west do not attract me. I will tour through South American republics as a respected engineer for five or six years in order to satisfy my impulse concerning the reaction of nature to various human interventions. Then I will enter on a no less honorable career as a Privatdozent. I lack the inclination towards being a goldminer."

Obviously, he was really strongly convinced that he would head for South America, because he ended the draft of the letter to Professor Wittenbauer with the phrase: "Do give me the pleasure, please, Herr Prof., of some lines and I will gratify you with glowing descriptions from the Andes."

From August 11 to around August 20, 1912, Karl von Terzaghi visited the towns of Reno and Carson City in Nevada and, in California, Lake Tahoe, Eaglewood, Bijon, Glentrook, Fallen Leaf Falls, and Glen Alpine Springs, where he toured great distances on horseback. As usual he observed the landscape keenly and described it and the geological situation with eloquent words.

In Carson City he wrote the first draft of two letters to his acquaintance Dr. Josefine Kotzbeck, a friend of Ella's family in Graz, always called *Fini*: "I have already received two letters from you and I do not know which of them I appreciate more. In one case you console me after a grievous lost (Karl did not explain what he meant with 'grievous lost', the author), so lovely and sympathetically, as no other human being – in the other you pleased me with a lot of colors and sounds from home, just at a time when the memories of home are of a special kind ... My illusions go like the autumn leaves with the wind. I laugh after them. Everything that isn't nailed down should go, I do not go.

I do not dare to show the situation clearly to my people. Currently the future is grey in grey. I have already many inner struggles behind me" ... In several paragraphs in the draft of the letter he expressed some sentimental emotions when he spoke of his mother and grandfather. "I will finish my excursion in 6 to 8 weeks and will travel with the next freighter to South America, at first to Peru, and will thus destroy grandfather's hopes of seeing me again in the near future. Five, six years I shall wander as a hungry engineer through the South American republics in order to taste to the full down to the last drops the whole beauty and the whole roughness of the pioneer life, then I will take up the hard career of the Privatdozenten in order to shape all that I have collected, without hope of appreciation. If it pleases Fortune to garnish the meager food – all right. I do not demand it; the fact is enough for me that I haven't changed my skin under the most difficult circumstances ...

Today I will not write about Olga. The subject hurts me badly. Only one point. She complains that she is in hell. Mr. Ramos has lost his mind and his frantic tenderness and attentions pursue her minute by minute and her peace and quietness is gone. In addition, the revolution which makes correspondence with the outer world nearly impossible; a letter from San Francisco to Mexico takes more than two weeks and two weeks back.

With this it is enough with the horror. What I have described in this letter is the new world in the light – or better said, in the shadow of the sharp common sense. But of its subtile attractions, of the fairy-tale magic of the desert with its tornados and with lines trembling in the glow of the sunshine of the tropics and the puzzling fata morganas, of the unspeakable beauty of the nocturnal starry sky above the very old pines of the Colorado plateau, of lonesome rides, lasting for days, through the bush of Arizona and of the holy silence in the Sequoia primeval forests in the canyons of the Pacific coast, of the daredevil, boundless progressing of a new young natural generation on virgin ground with all its daring and rawness, with its weaknesses and its prospects, of all the possibilities with which the atmosphere so to speak is impregnated, I have not spoken, and if all of these pictures would unfold before your eyes, then the grey would disappear into a silent corner.–

With this assurance I will finish my epistle from Nevada."

During the next days Karl von Terzaghi visited several construction sites east of the Carson City area. At the end of August he rode to Salt Lake City. "Unusually firm streets, good shops, facades most sandstone or bricks and as tasteless as possible. Huge name-plates, on the lamp poles flags of the Union in pairs, firmly tied at the bottom because of the wind. On the temple square the powerful temple built with grey granite." In the following paragraphs of his diary from 1912 he described the temple in detail. Then he stated: "In the park two statues: Joseph and Hyrum Smith. Joseph Smith receiving the divine inspiration, a profile like Schiller, huge nose, and a fine sweeping, feminine mouth, with peculiar mixture of enthusiasm and coquetry, in a suit of the 1840s."

On August 26, 1912, Karl stayed in Provo, Utah where he went to the tailor, to the blacksmith and the photographer in order to repair his belongings. From Provo he rode back to Nevada, to the area between Hazen and

Fallon. Afterwards he viewed several other building sites in Nevada and took in the marvellous landscape of Utah on horseback.

At the turn of August to September, 1912, Karl von Terzaghi arrived at Portland, Oregon. He only stayed here for a while and then rushed to Yellowstone, located in the heart of the Rocky Mountains. This oldest and largest National Park with its many geysers and sinter formations, the grandiose canyon of the Yellowstone River, the waterfalls, the great Yellowstone lake and last but not least the many animals living here, is without a doubt one of the most famous national parks in the world. It was bitter cold, when Karl arrived and the tents and log-cabins were heated. He hiked first thing in the morning through the wonderful places and in his diary he laid down again a real travel guide of Yellowstone.

In the middle of September, Karl von Terzaghi was in Rupert, Idaho. He visited construction sites here and in other places in Idaho, where he discussed construction problems with civil engineers and did a lot of sight-seeing.

At the end of September Karl von Terzaghi followed the Snake river with its huge water mass and waterfalls. "The Shoshoue Falls, this masterpiece of nature, this jewel in dark basalt frame, lie in the heart of a young, flourishing irrigation district", close to Twin Falls, Idaho. Northwest of Twin Falls Karl viewed the impressive construction site *Arrowrock* and on his 29th birthday he left this area by train and rushed for Seattle via Sunnyside and Yakima. One evening when he sat in the Hofbräu Café and listened to a German band playing old, sentimental songs, he got very depressed. "All the people who I have loved, who are connected with me by heart's blood, emerged unsummoned from the oblivion, my solitude came dreadfully to my awareness ... My fate carries on in this way step by step. What all fascinated me in earlier years? ..., botany of the Kryptogamen[23], theory of cognition, and astronomy. The secret fields of chemistry and of physics of the ether. The complicated problems of tectonic geology. Each field includes in itself lifetime tasks. All these interests must vanish or must be restrained within harmless limits in favour of the only thing. Combat of the natural powers, mastering of earth and rock work. All, other things must vanish, if I am to bring it to a head in one field and escape the danger of dilettantism, the awful. What choice do I have? Hydraulics professor who has never built anything – a caricature. Geologist? – platonic joy for retirees. Office-holder – and all my urges and intentions condensed to a subscriber of a rental library? No. I must disclaim all the tidbits of the spirit and go into the most concrete form of all concrete forms. What I have always asked for is present in this field in some form: natural science. – Each drilling opens a new view into the belly of the earth and the right answer comes upon each irrigation. – Beauty – Isn't the miner between rigid mushroom-like timber, the ripping party at the setting up of a crane, the working steam paddle, beauty enough? ... Philosophy and science

[23] Fern, braid, moss

of social science as welcomed bonus. Grit your teeth and let's go, this is all that I can do and must do, otherwise I become a dilettante in the profession."

At the beginning of October, 1912, he wrote in his diary that he had decided to go to Peru, where he was attracted by the huge mountains. "It is just the calling of my life to develop all the skills which I possess as completely as possible. I have a certain hesitation going back to Europe, even for a short time. Europe is the land of the sins of my youth. There I developed, alongside many good things, all the bad seeds in my nature."

Back in Portland (October 10, 1912) Karl von Terzaghi received a letter from his grandfather and he was deeply shaken. Obviously he had written to Graz that he would like to manage grandfather's own funds. Earlier in his diary he had indicated this possibility. He stated there that he would dispose of his grandfather's fortune during his travel through South America. Grandfather rejected this wish in his letter, obviously very brusquely. "Truth often hurts and this letter contained a lot of this kind. I had hit grandfather's most sensitive spot with my San Francisco letter. However, grandfather's reply should remain law for me concerning his finances despite its hardness and despite the injustices which it contained. I do not have the right to dispose of grandfather's fortune in another way as he wants neither before nor after his death. Besides I am forming a union with the engineering professions so tenderly that an investment, simply tending to a speculative gain, would run counter to my personal interests. Grandfather's emotional state, which is expressed in this letter, makes my trip to Europe necessary. A heart-to-heart talk can bring about, so I hope, something good." For the first time Karl expressed the possibility of going back to Europe and not rushing to South America. On the next day he wrote a letter to his grandfather. "It is strange in life. We build quietly and slowly our plans, consider this building for the decisive things and then – fate comes like a storm, and destroys all that which is edifying and speaks: stay, it is so, not the other way, and we are brought forward by an impact, however, in a completely different direction than we expected."

The rebuke of his grandfather had hurt Karl von Terzaghi deeply. As usual, when somebody was not of his opinion and he could not enforce his own will, he accused his opponent, offended him and tried to put him down. So also his grandfather who had done so much for him. In a diary entry of October 15, 1912, written in The Dallas, Oregon, east of Portland at the Columbia river, he pointed out: "It is sad that just now so hard, opposed, indignant feelings must urge me against him, immediately after his 90th birthday (it was the 89th, the author). However, these feelings must be expressed, otherwise I have to apply force to myself and the truth. He had enjoyed the so-called repose for forty years and had terrorized his family; without knowing it he has granted himself a quiet sad eve of his life.'All human beings are beasts.' This is his last resort. The thought of mother's inconsolable existence is dreadful for me and also his self-worrying and bitterness. The cause for all that became clear to me today, when on the steamer I had to again and again think over all the things which bother me against my will. He is, without any doubt, a man of high intelligence and iron energy, uncompromising stubbornness. He has constructed his lifework with fullest consequence and with keen logic, stop – his lifework? In the last moment, at the zenith of his successes he has made an error and he relies heavily on this error today: he considered the creation of his prosperity as his lifework and not work which has satisfied his life. He observed the conversation, the wise pleasure of life and the useful application of this prosperity as the continuation, no, as the crowning moment of his lifework and devoted to it, in the time to follow, all his years. And look, today he stands shocked before the bankruptcy of his intentions, his wishes, his hopes. And acrid bitterness accompanied each of his talents, each of his thoughts. Instead of being absorbed in the further interests of his profession, in the philosophical working up of his experiences, after the ending of his lifework proper and to tie up through threads, and always again new threads, with his time and with nature, he took up the obstinate, selfish pursuit of his pure personal intentions. And on that the happiness of his old age has failed: instead of accomplishment he sees contradiction, instead of the expected thanks, mistrust. It is sad to see how a man of excellent merits and excellent abilities steps as a shadow into well-deserved retirement. Poor mother, the only creature who stays continuously at his side, appears as exhausted by his eternal gall, bloodless, nearly ghostly and deprived of her personality. From time to time she weeps quietly inside herself, and then she bears it again in desolate devotion, the yoke of her bleak life!

Protest against each approaching life destroying power, and would this protest appear even as a shabby ingratitude. Money is something, however, life is everything, and no money sacrifice in the world can make up for an attack against the right of life.

If grandfather would have dedicated his time during his retirement to work at the writing desk, to studies on the progress of the tobacco industry, to the state-economical importance of monopolies, to working out proposals, than he would have kept himself lively and would have developed his increasing knowledge as an alternative to all of the fiascos which occupy him exclusively today."

It seems that Karl's grandfather had, in addition, stopped the money transfers to America. Otherwise, one cannot understand the rude, unjustified hate tirades[24].

The weakness of character of reacting to alleged opposition to his own opinion, his will, was not on account of his youth but rather he kept this weakness his whole life. Decades later he offended his life-time disciple,

[24] One may guess that his grandfather had disinherited him. Not only these hate tirades point to this possibility but also the fact that he was sometimes in real money trouble in Constantinople after the death of his grandfather in 1916, although grandfather had promised him a great amount of money as a distributive share.

A. Casagrande in a very rude tone when Casagrande asked Karl about writing his autobiography. Karl understood this wish as an affront against his ability to continue his scientific investigations and he took revenge, casting doubt upon Casagrande's competence for doing research work.

On October 18, in the afternoon, Karl von Terzaghi headed for San Francisco with the steamer *Beaver*. Three days later he arrived at Golden Gate and San Francisco. The purpose of his voyage to San Francisco becomes evident in a letter to his respected friend in Graz, Professor Wittenbauer: "At the end of October the long row of western and wild-western hydraulic constructions had been passed and I came to San Francisco with the insight that a concluding judgment about the value or lack thereof of new working methods, unknown to me, cannot be formed from observations and optimistically coloured information alone. I offered my services. Received offers from construction bureaus, however, none for buildings work. Put therefore my academic education into my pocket and went into service as a drillman at a building site of a government lock at the Columbia river with a large collection of heavy drilling machines, steam shovels, sand excavators, ... A hard and dangerous business."

In his diary he struggled for the right arguments justifying his decision to work as a drillman. He, as an academically educated young man with a doctor's title went into the low points in life. Not only the fact that he had to do really hard physical work, but he also had to live in a rough society of adventurers, lumberjacks and fortune-hunters. As usual, he turned away from his decision by discussing, in an abstruse manner on many pages, the education of engineers at the Technische Hochschule and concluding that the professors neglect the instruction in the practical side of the engineering profession and focus too much on the theoretical field. He came, finally, to the realization: "The value of the engineer lies therefore mainly in the field of the right selection of the machine stock, in the field of organization and, in particular, in that of the estimate. All of these three sides of engineering are neglected at the universities, though they, however, represent the most important things."

At the beginning of November, 1912, he started his heavy work at the construction site of Celilo Locks close to the Big Eddy rapids. After the first very hard days at the building site he had adjusted to the situation. "Strange, how each seemingly unbearable situation can be born not only with patience and dignity but even with rich gain and how unusual circumstances produce unusual knowledge. I will not forget with which feelings I accepted the laconic offer of Mr. ... in front of the door of the office building on a rainy Thursday afternoon, whether I would carry drilling tools from the pit to the forge? He does not have other work and still he would give me the opportunity to learn the steam drill business. The way from Camp 1 to Camp 3, eight kilometers long, was indeed a walk to Canossa; however, this information with the accompanying idea of crowded quarters and rough treatment put me in a real gulf of consideration and my *yes* was obviously not very convincing. However, and this is the main thing, it was said and thus I swam. During the night from Thursday to Friday I slept already in the small wooden partition ... and in real surges of snoring.

The first Sunday saw for me a deadbeat, with sore limbs and serious doubt whether the next (Sunday) will still find me alive in Big Eddy. Today, on the second, I am not only lively but I have ... found again my familiar humor and I feel like a winner on this battle field of basalt and steam. The power of fantasy is still wonderful." And some days later he stated: "No day without incident, a merry, hard combat from sunrise until deep into dawn. And each day I feel stronger, freer and healthier. My dissolute life in St. Petersburg appears to me like a chaotic dream. It is really strange. I had to travel through half of the world. With restless, unsatisfied, eternal doubting, eternal grumbling spirit from the dolines[25] of the karst shadowed by beeches and from the churches of Rome to the Archipelago[26] of Finland, from Mexico to Canada through deserts, through canyons and over mountains, dissatisfied and searched like Carlyle's Mister Teufelschreckh (?), in order to find, finally, on the Columbia between basalt and steam the courage to throw everything from myself which does not belong to me and to be myself."

The work at the building site of Big Eddy was not only hard but also extremely dangerous. At the beginning of December, Karl scalded his foot, which was very painful, and over the next days he spent the evenings in bed because he suffered also of dramatic fever. In the middle of December the pain was so strong that he could not sleep. Finally he visited the doctor who stated: "Three days later and your foot would have been lost." On December 21, he left Big Eddy with a little sweet melancholy and spent the rainy Christmas holidays in Portland. A special chapter had come to an end as a result of his injured foot – and his restless nature as he indicated earlier in his diary. Life in Portland appeared to him bleakly sober and boring. In this dull mood he saw a note in the *Morning Oregonian* of December 29, 1912, about the death of a hobo, Max von Bülow, dismembered by a train, close to Reno, Nevada. "This passion I cannot conceive. The passion of the 'beating of the way', the diabolical magic power of the 'call of the rail'. ... The effects of the passion are more shameful and more destructive than those of boozing. From one jail to the other, despised and stigmatized. And there seemingly normal human beings left their quiet, comfortable home, irresistibly attracted by the single thought of the 'call of the rails'. And run if they are brought back again into the civil life. Are they 'would-be frontier men', who desperately hoping, but confidently pass through the land to reach the paradise of their dreams? Has the wandering Jew possessed them? Do they run into the terrible bleakness of their own insides? It is a psychological puzzle." He digested the desperate

[25] Dolines are funnel-shaped cavities in the karst.
[26] Small rock islands and coast cliffs of the Skandinavian and Finnish coast.

life of this hobo, who carried a noble name as he did, in a feature about vagabonds. It seems that Karl was anxious about suffering a similar fate.

Karl von Terzaghi's deep depression, which was, without doubt, intensified by an unhappy letter from Olga Byloff from the beginning of January, 1913, his great self-doubts and his complete solitude lasted until the end of January, 1913. On January 24, he confided to his diary: "After a row of days in deepest depression, in discouragement and self-torture, I'm as if transformed. I was free only for two possibilities: to double the self-confidence or to collapse untenably. The feeling to intend to work and inability to do this, the feeling of unemployment, is dreadful and it requires much self-confidence in order to keep the head up and to look around naturally as if all things are in order ..., the self-loathing leads to alcohol and the shipwreck is ready." Karl tried to overcome his depressed mood by writing essays, attending lectures and visiting theater performances. In this period he wrote nearly twenty non-technical essays. The subjects of these essays were, in parts, very different, from "Die neue Demokratie" (The New Democracy) to America's commercialism. His essays remained mostly unpublished. Although they contained some valuable thoughts, his treatment of the different subjects was unsatisfactory in some respects; the treatises were too wordy and humorless; they were sometimes constructed in such a way that some parts were described in full while others were only brief, and sometimes his train of thought got lost. Moreover, it seems that the range of themes were uninteresting for many publishers.

On Wednesday, January 22, 1913, Karl von Terzaghi attended a lecture about Ibsen's *Der Volksfeind* (Enemy of Society). He was not very amused about the talk; "a dry, moralizing and sometimes ordinary pathetic statement of contents with invectives against the 'saloon' character and against the restlessly acting vampire of society 'and we are dreaming, dreaming, dreaming, and once when we wake up, it will be too late!'."

In the evening he visited the theater performance *Peter Grimm's Return* which was really enjoyable for him. "Convincing representation of the unconscious relation between a deceased man and men of his old surroundings ... Lack of discipline of the audience. How primitive the theater is."

In March, 1913, he saw the unforgetable actress Sarah Bernhardt in a performance of the *Lady of the Camelias* and was very impressed and moved by her acting and later disgusted by her professionalism. "Sarah Bernhardt is old and has come down from the theater of her victories, the Théater Français to the vaudeville theater. Nevertheless, who can withstand the power of her acting. That was not only an actress who acted for us the death of the lady of the camelias; this was a dying person who suffered unspeakably before our eyes, in whom the fading life desperately struggled with the undiminished love-passion of the yearning and desiring woman ... When then, after the fascinating unifying-scene the unavoidable occurred, the accident must hit actress and spectator with the same hardness.–

However, no. One minute had barely gone and the curtain raised itself again and on the stage a female monster appeared with a squeaky voice and sickening coquetry and quacked saying: 'he was nice – he was nice – he was nice – he was nice – but he wasn't the man for me' and without protest and without indignation like a herd of stupid domestic animals the audience remained sitting and chewed further. I ripped up my program in small pieces in rage and went away as fast as I could."

Despite all his free-time activities he still remained down-hearted. Unemployed, his relation to his grandfather strained and far away from Olga Byloff he was barely making a living. He recognized that he had also failed to solve the task which he had formulated at the beginning of his excursion through the Southwest, West and Northwest, namely to create unique relations between the geological character and the building-technical behavior of soil and rocks. Moreover, it seems that his grandfather had interrupted his remittances and that Karl was forced to take every kind of work to earn money. On March 6,1913, he expressed his worsening situation in a letter to Professor Wittenbauer: "I am experiencing the most miserable period of my entire life so far; unemployed and my mind divided and tortured by doubts. Sadly I have to confess that thoughts of home fill me with indignation, far away at home they grieve over my existence in a selfish, small-minded way, and my communication with them is limited to dry, factual reports. How mean, unimportant the security of my existence is when I have lost what I most valued, my belief that engineering is the most blessed profession, and the prevailing hypocrisy fills me with disgust. Even love for a child cannot hold up against these feelings. I have broken off all correspondence, with the exception of that with my dear friend, Mrs. Dr. Kotzbeck, because I do not wish to involve others in my depressed state of mind, and I wish to avoid the pain inflicted by an indifferent shrug of the shoulders, with which even old friends are brushed off under such circumstances. I have no one here, only the stubborn decision to keep going, and not to leave the United States until I have regained my old confidence in a new form."

In March, 1913, he also interrupted his entries in his diary and he took them up again when he was already on his voyage home in December, 1913.

His spirits were raised again from the depths when he became acquainted in a tavern with an architect who was having difficulties with the structural design of a church cupola. Von Terzaghi took over the job as a designer because he was familiar with the problem through his dissertation. Moreover, through the negotiation of the architect he got the position of an engineer for the foundation of large buildings for the Portland Gas and Coke Company on the Willamette river. Furthermore Karl von Terzaghi learnt the details and mechanics of soil excavation using powerful water hoses widely used in the United States.

He expressed his rising spirits in a letter of November 5, 1913, to Professor Wittenbauer: "Today my courage is again as fresh and dashing as that when

I was twenty-two years old. What at that time had been alive concerning good things is today even more alive." And furthermore: "..., that I have also won through here, in a hard country and in hard times, have learnt from friends and foes and that I will leave the country as a weaker, more righteous, prudential man than when I was placed in this country two years ago." Finally, he announced his wish to come to Europe. "That I will visit Europe before I disappear to South-America, my brother-in-law has probably already told you."

How his second wife Ruth saw his extended stay in the United States can be read from her letter of December 13, 1989, to the author: "Following his work with the US Bureau of Reclamation described by A.C.[27], he spent some time in our southwest, where he even briefly considered setting up in business as a date farmer (I got this story from Ella, to my great surprise, as he never seemed to have any interest in agriculture of any kind). Later, he went to Mexico, put money into a construction business and lost everything as a result of the Mexican revolution. Next came the northwest and his work as a driller, followed by a job as a designer for either an architectural or construction firm."

The hint that Karl lost all his money may be the cause for the deep discord with his grandfather. It may also be a hint that he had visited Olga Byloff during the summer of 1912 as promised in New York.

On December 9, 1913, Karl von Terzaghi left the United States on board the relatively small Austro-American steamship *Argentina*, a name which contained a slight irony. There were only four people travelling first class and, with Dr. M., a German ship's doctor, he developed a close relationship and he saw in him a piece of home. The voyage to Triest required three weeks. As a result he had time and leisure to think over economics, science and ethics. He wrote down, partly on board, partly in Graz, his pseudo-philosophical ideas in his diary. On New Year, 1914, he was back in Graz. His efforts, during the next month, to apply the great experience gained in the US in Argentinien,

---

[27] Arthur Casagrande was born in Haidenschaft (South Tyrol, Austria) on August, 1902. After he had passed the Elementary and completed the Real School in Linz, Triest, and Vienna he entered the Technische Hochschule in Vienna in 1919 in order to study civil engineering. During his last year in the Technische Hochschule he was an auxiliary assistant to Professor Schaffernak and after his graduation his assistant. The bad financial conditions at the Technische Hochschule forced him to go to the United States where he arrived on April, 1926. He obtained a position as draftsman for Carnegie Steel near Newark, New Jersey. Soon he met Karl von Terzaghi and became his assistant at M.I.T.

In 1929, he went with von Terzaghi to Vienna to equip a soil mechanics laboratory and returned to M.I.T. in 1930. In 1931 he was offered a one-half-time lectureship at Harvard and installed an effective teaching program during the next years. In 1934 he was promoted to Assistant Professor, in 1940 to Associate Professor and in 1946 he became Gordon McKay Professor of Soil Mechanics.

Arthur Casagrande was active in many field of geotechnique. He published more than 70 papers and educated highly qualified persons for teaching, research and practice.

Karl von Terzaghi
in Portland, Oregon, 1913

Karl von Terzaghi
at age 29

failed because of a severe critical business depression. Also his plan to found a construction firm with two professional colleagues was foiled by the outbreak of World War I.

He continued his pseudo-philosophical considerations, in particular on the outbreaking war and related problems also in St. Pölten, where he was obviously on vacation, in August and in Vienna Neustadt in November and December. He lived there in the house of Olga's parents. One can only guess that the relationship to his grandfather was still strained and that he stayed in Vienna Neustadt for this reason. One does not learn any hint from his diary neither from his daily life, and neither his relationship to his family in Graz nor of Olga Byloff's fate.

After his bad experiences in the last months of his stay in North America and his return to Europe at the end of 1913, Karl von Terzaghi was pleased to come back to the world of old Austria, although soon this old Austria would be gone and he would be forced to face new challenges.

The old Austria, the Habsburg monarchy, was a remnant from the past, the only non-national state in the era of nationalism (Mann, 1958). The political collection of landscapes and different people named Austria was however, not created by pure chance. One could live pretty well in old Austria during the last decades under the reign of Emperor Franz Joseph. There was increasing prosperity and legal unity. The culture was merely German, though a special kind of German, opened to the influences of the south and southeast

The arrested assassin Gavrilo Prinçip

and fed by old cosmopolitan tradition. Besides that, the striving of the small ethnic groups for self-realization had been growing steadily: the Czechs, the Poles, the Hungarians and the Croats. Enthroned above his people, the Emperor, a man from a legendary past, the young man of the counterrevolution from 1848, had become old, experienced and pessimistic. The people loved or admired him for his strict life-style, his dignified appearance and his loyal work. And the people liked Vienna: *Kaiserwalzer* and *An der schönen blauen Donau*, the officers and the soldiers of the k.u.k[28]army, Hofräte and Kommerzialräte, Burgtheater and opera, the poems of Hugo von Hofmannsthal, rich, noble and heavily intellectual, the dramas of Arthur Schnitzler, the famous scientists in medicine and physics. How could this world not last? Why did the disaster start just here, which changed Europe so deeply? And it happened soon, although the Austrian state had many liberal and tolerant characteristics. However, there was also a lot of hate in Vienna, in the surroundings and in other parts of old Austria. Hate of many people against Jews, hate of the bourgeoisie towards the social-democrats, hate between the Germans and the Slavs on either side, mutual hate between losers and the rich.

The German nationalism in its milder form wanted to keep the German character of the whole state and to strengthen it; in its sharper form it was prepared to disintegrate the monarchy and to seek an annexation by the German Reich. However, the internal conflicts were not so strong as to change

[28] Kaiserlich und königlich (imperial and royal)

the situation in Austria decisively. It was her foreign policy which finally brought old Austria to an end.

On June 28, 1914, the Austrian successor to the throne and his wife were assassinated in Sarajevo by south-slavic nationalists. Austria blamed Serbia for the assassination and sent a sharp ultimatum to the Serbs, after Germany had ensured its complete support for Austria's actions. Serbia accepted the main part of the ultimatum. It declined, however, the main condition, namely, to allow Austrian authorities to investigate the background of the assassination in Serbia. The following Austro-Hungarian declaration of war on Serbia on July 28 caused Russia to take Serbia's side and to start mobilization. This, in turn set off over-hasty diplomatic and military steps in Germany, which declared war on Russia and France at the beginning of August, 1914. The international system of alliances quickly caused one declaration of war after the other. Germany, Austria-Hungary, Turkey (November, 1914) and Bulgaria (October, 1915) joined to form the so-called *Central Powers.* On the other side the *Entente* joined, Great Britain, Japan, Italy, France, Russia, from 1917 the United States, and many other countries.

Already on December 2, 1914, Austrian soldiers were successful on the Balkan front. They occupied Belgrade. However, the Serbs pushed back the Austro-Hungarian army behind the Danube and Save in the middle of December. Austria-Hungary lost 227,000 men (dead, injured and captured) out of their 450,000 men in combat against the Serbs. The military failures hurt the reputation of Austria-Hungary considerably. First by a German-Austrian-Bulgarian attack the Central Powers were successful and captured Belgrade in October, 1915, and subsequently Serbia, Montenegro and Albania. In October, 1915, troops from Great Britain and France landed in Saloniki and opened a new front in Macedonia. Turkey could not support Austria-Hungary very much. It had lasting success only in the Caucasus Mountains against Russia. After victories at the beginning of the war in Iraq against British troops, the Turkish troops had to retreat and the British soldiers captured Baghdad. Moreover, in 1917/18 Palestine and Syria were occupied. However, British-French breakthrough-attempts in the Dardonelles were in vain. After the armistice of Mudros in 1918, Greek troops occupied Smyrna.

Karl von Terzaghi got involved in World War I as a first lieutenant at the beginning of January, 1915.

On December 22, 1914, his draft papers were sent to his Vienna Neustadt address with the order to come to Vienna immediately. On January 6, 1915, he assumed his duties in Dreher's beerhall in Vienna, after he had put together his luggage for the campaign. He had to form a Landsturm (veteran reserve) battalion charged with building trenches, fortifications, roadways and bridges at the Serbian front in front of Belgrade. "My experiences of many years in treating and organizing masses of workers already helped me in Dreher's beerhall. Immediately, I put the unruly or intoxicated elements in their place, forced the active troops to perform their duty with the same formality and the same punctuality as a company in readiness for march and called the veteran reserve's attention with special emphasis to the war article after swearing in the court of Dreher's beerhall. I succeeded in bringing the company of 260 men to the quarters in good order, ..." Moreover, shortly before the departure to the front, Karl von Terzaghi carried out all the invalids by himself, because there was no doctor available. In the midst of January, 1915, he left with 4 companies (about 1,000 men) from the railway station Wiener Neustadt, Vienna and headed with the train (32 wagons for the workers, 1 wagon for the officers) to Hungary. The ride was very strenuous and in parts badly organized. In Ujvidék (Novi Sad), for example, the companies had to leave the train despite Karl von Terzaghi's protest. Quarters were not available. He had to lead the soldiers to a free place where they had to wait for many hours in foot-deep mud, dense snow flurries and darkness for further orders. Many of the people had worn clothing and torn shoes; several were suffering from exposure and other diseases already due to the strenuous transport. However, the administration in the base held office first from 9 a. m. Karl von Terzaghi started his way through the different offices at 9 o'clock. Obviously, nobody was responsible for Karl von Terzaghi's companies. Finally the chief of the ambulance men, an old, energetic and intelligent doctor in the general staff, gave the permission for quartering in two available shacks which had space for at the most 200 men. However, Karl von Terzaghi organized 1,400 kg of straw, which was sufficient for one thousand men to sleep in the shacks. He ordered the quartering of five hundred men in different houses. At the railway station he became acquainted with a commander of six hundred men who were also without quarters. They had arrived from St. Pölten one day before in the evening in a very exhausted state after a ride of five days. The commander was obviously overburdened and Karl requested him not to look further after the business of his companies. "I had taken over the command." In the evening, Karl made an appeal to the men in the two shacks to make room for the comrades in mud and darkness despite the lack of space. His men agreed and one hour later all workers lay crowded together on the straw. After one further day, Karl and his men rode with a steamer on the Danube to the south.

Before the departure of the steamer, Karl von Terzaghi became acquainted with a captain who complained bitterly of the state of the south-east theater of war. The confusion in the administrative departments was unbelievable. One command contradicted the other. The worst institution was the general staff, with its purely theoretical education and lack of experience.

The ride with the steamer went perfectly. The weather was clear and sunny. The workers arrived at their destination Old Slankamen in good form; a part remained there and a larger part marched to New Slankamen. On the main street there was mud up to half a meter deep. The inhabitants with their jackets of sheepskin and high fur caps reminded him of darkest Russia. Also for the troops the hygenic situation was worse. The wells lay between dunghills. Toilets were nearly unknown and the kitchen for the troops was located in a house where typhoid-convalescents stayed. The responsible doctor was not only a lazy man who was not interested in the well-being of his patients but it seemed that he could not handle the situation in factual respect.

Despite all the adverse circumstances the excavation of the trenches around New Slankamen made good progress and was nearly finished by the midst of February, 1915.

In this chaotic and dirty world Karl von Terzaghi found leisure time to reflect on his situation and to think about the war. He remarked that there was no pure coincidence and that the balance of fortune in our life could not be influenced by the merits and efforts of other people, and he continued with deep thoughts: "Considered from this standpoint, war appears to me as a grandiose, mysterious occurrence; like a Last Judgement, in which mankind, put into a bloodlust by invisible powers, executes a mass judgement of fate on itself in a bloody manner, forgetting all gained insights and against all worldly wisdom. When already an accident, which comes as a surprise over a naively living human being, has to inspire us with awed shuddering in the presence of the hidden hand of God, how should the spectacle grip us, which the unleashed powers have violently attained over mankind and destroy hundreds of thousands and hundreds of thousands of existences in the course of a few months. However, people only see and do not understand. They do speak of a 'fateful hour of their empires', but they think war is the result of reckless diplomatic intrigues and would have been to abandon. They botch the most valuable opportunity for insight into the being of our existence through their superficiality.

While we consider ourselves as planning and acting for Providence and believe ourselves to be able to lead the fates of human beings, we only span the bow; the darts are laid in order by fate. It is good to familiarize oneself with the thought that one is only to a very limited degree master of his own fate, to say nothing of master of the fates of others ...

A poetic tinge sticks to the war, coming from the time when it was the business of a selected number of daring, crude, merry and energetic fellows, a tinge which it does not possess anymore. War has become a science which is practically tested only every half century and shares with the theoretical sciences the fate that it can be operated by bunglers but that the public can recognize the bungling. Something of dillettantism sticks even to the considerations and the actions of the gifted just as to the projects of theoretically trained engineers, because the thinking of these people has not passed the crucial test of experience. Numberless heads swim in confusion if fateful seriousness emerges from the theoretical science of war; many, many men, called to acting and leading, must bow down before the artificial-made authorities of incompetent theorists, see their fate in the hands of people with insufficient judgement and suffer heavily from obstructions. Only seldom do the situations find men who can stand them; on each gram of fresh daring deed, many hundred-weight of stupid grey misery drops. The people in the hinterland yack about heroism, indescribable spirit and of inspired moods without knowing that the requirements for the unfolding of real martial virtues are only seldom available."

In Karl von Terzaghi's statements there are two remarkable points: firstly, it is amazing how clearly he saw the huge war development from a more romantic adventure with relatively few victims to the modern mass destruction based on the industrial revolution. Secondly, already in 1915 he expressed his deep disgust for all theory not based on observation and experience. We will see that he repeated his standpoint in the following years again and again.

In a first draft of a letter to a certain L. K. from April, 1915, he stated further: "My passionate efforts coming from the heart were applied to dethrone technique and natural science during all these years, at first inside myself and then also outside myself, ... I work in the service of progress and I make an effort to prove what I feel, namely that the idea of progress is a fatally, criminally misused wild notion. And in this way it happens to me with all things. I need, as a weak, imperfect human being, the small successes in accepted areas of work in order to keep my self-confidence for such problems which I work on without support – yes, without hope of understanding. Possibly, within ten years I can publish a consistent version – with a treatise 'Old and New World, a talk on culture' I have already put down in writing the core of my views and the outlines of my program – I do not know whether I gave it to you to read this small, though weighty, writing at that time. If you realize how high-handed, yes, how absurd the standards are which we use to measure the value of the acquisitions of our time; how far the view for proportions in progress-trash is lost, you can realize how much I have to do ... I would like to state with certainty that the publication of your research results has brought to you only bitterness. With mockery one would put you to the side. You do not know the rabble of future scholars as exactly as I do. The stamp of the impetus of research, which you wear on the forehead, is regarded by many eyes as a mark of Cain and a reminder of his career as a dangerous peace breaker. He who writes because of the truth turns to half a dozen rare human beings."

On May 24, 1915, Karl von Terzaghi received a telegram with the command to prepare his department for an evacuation with a steamer which was to bring them to another place. He made the marching preparations and in the morning of the next day the column moved on with flying colors, with music and unavoidable drunken workers and was accommodated on two barges. Soon, afterwards they left their working area for fortification construction work in the Aradi region (Hungary), north of New Slankamen at the Tisa.

Gradually the service in the veteran reserve was going to become boring for Karl. At the beginning of August he requested his friend Hans Kalbacher to make it possible to transfer him to the aircraft department in Aspern, close to Vienna. His wish would become reality very soon.

On August 9, 1915, Olga Byloff came to spend several days with Karl: "... with her row of sunny days and happy hours of heartfelt friendliness. I hiked together with her through meadows and swamps in the surroundings and was glad for the fond, unselfish love which the strange girl always pushed again into my nearness." On August 17, they had to separate again with heavy hearts.

During Olga's stay, Karl von Terzaghi received a letter from Hofrat F. (Forchheimer) in which he was asked whether he would accept a professorship in K. (Constantinople). With great pleasure he thanked him for his goodwill and asked Hofrat F. to arrange an appointment.

At the end of August, 1915, Karl von Terzaghi watched the marching off of a German battalion to the North. The officers of the Austrian army accompanied their German comrades a while and said fare-well with emotion. Amazingly, Karl, as an Austrian officer, praised the Germans, the unbelievalbe qualification and the exemplary organization of armed Germany. "After the first glance which I ... gained in the war area the defects in our army administration appeared to me doubly inconsolable." Works on fortification proceeded with changing speed. Many delays were caused by the eternal disparities of views and petty jealousy between two majors. One of the majors was a notorious alcoholic and generated bad blood within the worker's ranks through his unmotivated outbursts and threats. Finally, this major received, to Karl's greatest surprise, the command to go to another town – and the peace was saved.

The following weeks were filled with many social activities, some parties and visits to acquaintances.

At the very beginning of September, 1915, Karl von Terzaghi and his workers were transferred to the inner part of Hungary. However, on September 5, Karl rode back to the area south of New Slankamen, to the front line in the northern part of Belgrade in order to fix some administrative problems. On the way he saw the damage which the war had already caused. On September 21, 1915, he rode a horse north of the front line and came into cannonfire, in which his horse was wounded. At this time Karl von Terzaghi and his veteran reserve workers were mainly occupied with repairing of roads. They met endless columns of German infantry and several military trains with young German soldiers. It seemed that the offensive operation against Belgrade was to occur soon. It was expected that this would happen around September 24 and 25; however, it did not. The reasons were unknown.

On October 1, 1915, Karl von Terzaghi and his men were ordered to an ill-reputed Zigeuner (gypsy) island (Ada Ciganlija) which had become infamous in 1914. On December 2, 1914, the Austrian army had occupied Belgrade, however, it had to retreat on account of a Serbian offensive beyond the Danuva. In particular, the battle on the Zigeuner (gypsy) island cost nearly 8,000 soldiers their lives.

Finally, as nearly all had hoped, on October 5, 1915, at 1:45 a.m. heavy gun-fire set in lasting until October 8. In the evening of the next day Karl observed garish lightening behind the plateau rim of Belgrade. A red reflection of fire spread above the deeply hanging clouds and half an hour later the rim was in flames – the battle for Belgrade was taking its course. Karl also saw crowded military wagons with unruly officers, some of them bandaged, blood spots on their dirty coats. Moreover he observed Serbian prisoners of war. "Young people with pale, hollow-cheeked faces, dirty and unkempt, apathetic and exhausted." After the artillery bombardment the united Austrian and German troops captured Belgrade. Karl reminisced in a letter to his son decades later: "The siege of Belgrade and the capture of the city on October 9, 1915, were spectacular military events I shall never forget. For three days and nights the heavy guns sputtered fire and steel and paved the way for the final assault."

At dawn on October 11, Karl and his workers moved along the destroyed Save dam fortifications and over the new pont bridge to the Zigeuner (gypsy) island. The infantry had captured the island with heavy losses. The task for Karl was to build a firm road across the swampy island in a hurry. The town Belgrade was crowded with marching columns of German infantry. Near Belgrade an endless train of munition columns, marching columns, provision squadrons and combat engineer departments moved slowly to the war bridge. On the slopes, south of the Zigeuner (gypsy) island, one shell after the other exploded. When Karl von Terzaghi landed on the Zigeuner (gypsy) island it was already a bright day. "Traces of the embittered fight all over. Shells have dug deep craters in the wet soil of the virgin forest. Dug-outs thrown in, backstops destroyed. Bloody pieces of clothing, smashed rifles, ammution cases, bread sacks, canteens, ammunition boxes and cans between paper scraps and human excrement. A picture of inconsolable devastation. A communication trench, half hidden in the thicket and the forest twilight. Swamp green water filled it to the rim. And in the water lay the dirty, clay-yellow dead bodies of Serbian infantry; irregularly side by side and in confusion. Someone crouched with deep bowed head and crossed arms in the trench; some full bullet magazines lay across his broad back. Another lay on his back and it seemed that he stared up at the canopy of leaves, which hindered the view to the sky. A third person, a man with fine, sharply-cut features, seemed to be sleeping on the dead body of a fallen comrade. Most of them rested like dirty bundles of clothes on the bottom of the trench and were covered with clay which had flooded in. Here and there an arm or a leg rose out of the water. In the twilight, between the yellow wicker work and blackberry vines, the deep green, still water surface looked like the horizontal glass plate of a museum case, in which one has put on display the horrible picture as an edification and advice for all young men who enthusiastically sing in the hinterland odes to the onset of war."

Oberleutnant Karl von Terzaghi, 1916

Karl von Terzaghi and his workers began immediately realizing their task despite gun fire which seemed to be coming closer. Under a willow tree Karl found the dead bodies of six German infantry soldiers. He removed the legitimation plates from the soldiers' chests, sat down on an ammunition box and soaked up the nightly picture. "The death may have come to one man while he was jumping. His legs were pulled up. A man with a brutal chin and with a wide, distorted face lay there with split skull. The right eye was missing and the helmet showed on the forehead a hole as big as a fist. A gentle young fellow, I believe his name was Hannemann and he came from Württemberg, seemed to be sleeping. The look on his face was peaceful, the eyes closed. At the beginning I thought of the contrast between the lust for action and the confidence with which the young people may have gone to the battle field, and their sad end in the wilderness of the Zigeuner (gypsy) island. I thought of the fright which would have gripped each individual if this picture had appeared as a vision on the day of marching out."

On October 28, 1915, Karl von Terzaghi was informed by a major that he had to go – owing to an urgent request from the war minister – to the aircraft department in Aspern, close to Vienna, in order to take over the command of the airfield. Already on the same evening he studied his marching route. On the next morning at 5:30 a.m. he gave his last department report and said good-bye to his men with a long speech. It was still dark and raining.

This was not parting from comrades, but rather withdrawal from a system of mutual mistrust, senseless business and fruitless efforts. The atmosphere of the theater of war had poisoned the relations between Karl von Terzaghi and his workers. Only a few of his men had recognized him and appreciated his work.

In the afternoon, Karl rode together with his replacement to Belgrade in a motor boat in order to introduce him to various authorities.

Two days later, he left Semlin (Zemun). In Ujvidék (Novi Sad) he interrupted his trip in order to say good-bye to some acquaintances. On the morning of October 31, he continued his ride to the north.

"I came to Vienna without feeling joy." The memory of all the misery, which he had seen at the theater of war weighed heavily upon him.

Over his first twelve days in the aircraft department he recognized gradually that the detail organization in the hinterland was as bad as it had been at the theater of war. The work in the aircraft department was marked by great incompetence and due to this the whole department was ineffective and in some parts also corrupt. Around the midst of November, 1915, Karl von Terzaghi was mainly involved in the design of a hanger tent; in his diary some calculations concerning some static problems and costs are recorded. At this time some famous professors were working in the aircraft department, namely Pröll, Hanover, von Mises, Strasbourg, von Kármán, Aachen, Kurtler, Vienna, and Schruttka, Brno; in particular, von Mises and von Kármán would

Officer Karl von Terzaghi, 1916

Karl von Terzaghi at the airfield in Aspern

have a decisive influence on mathematics, mechanics and engineering in the years to come. Also Fröhlich and the professor at the Technological Museum in Vienna, Fillunger, worked in Aspern. Fillunger was appointed director of the aircraft construction department. It was von Mises who advanced to the position of chief designer in Aspern. On November 18, 1915, Karl reported in a diary entry that von Mises and 12 technical designers had designed a huge warplane with airfoils 3 m thick and with a wing span of 22 m; the length was 14 m. The aircraft production in Austria in the First World War was very small, in no way comparable with that in Germany, and it seems that also von Mises' capital airplane was never manufactured. The main task of the engineers and other employees at Aspern was to patch up and test captured Allied military planes with the purpose of learning about any advances which the Allies might have made over their Central Powers counterparts. Karl soon became the commander of the test airfield of Aspern. It was a hazardous occupation.

The year 1916 brought four important events in Karl von Terzaghi's personal life. Soon after his arrival in Aspern Karl von Terzaghi learnt that Olga was again pregnant – from their rendezvous in Hungary. He could not repeat his behavior of 1912 when he sent Olga abroad. At that time Olga lived, obviously, in Vienna. And on May 13, 1916, Vera was born in Vienna. Later he described this event in a letter to his son: "In May 1916 my daughter Vera was born and a few months later I decided to marry her mother (June 7, 1916, the author). I knew it was a daring experiment, but I wanted to do my best to provide the child with a home." Karl failed to mention either the birth of Vera or his marriage with a single word in his diary.

Karl was deeply moved when he learnt that his grandfather had died at the age of ninety two, on August 2, 1916. This great man acting in loco parentis and as a mentor after the sudden death of Karl's father in 1890, had influenced decisive periods of Karl's life both materially and spiritually – and had left a significant inheritance.

While Karl was still busy with organizing the air force and the completion of the technical air field facilities, he received from the ministry of foreign affairs the order to take over the chair for foundation and road construction at the Kaiserlich Ottomanischen Ingenieur-Hochschule in Constantinople in September, 1916, at the recommendation of Hofrat Professor Forchheimer.

He wrote later: "On September 9, a dream from spring, 1914, became reality." The leadership of the test airfield was conferred to his friend, O. K. Fröhlich, his successor in the construction firm in Saint Petersburg.

1.4 The Entrance to Science

"The opposite of theory is not the practice but the observation, the experimental research. And the opposite of practice is not the theory, but the science which consists of theoretical and experimental research."

Paul Fillunger

"All knowledge and every increase in our knowledge doesn't end with a period, but rather with a question mark. An increase in knowledge means an increase in questions to be posed, and each of them is always replaced in turn by new questions."

Hermann Hesse

Immediately after receiving the offer he rode to the Bosporus with draft of the contract in his luggage. "Never in my life have I gone so straight into the unknown. I do not dare for once to predict whether or not Turkey will be my impending home. My fate is as uncertain as the future of Europe." However, at the railway station a problem arose. The commander, a captain informed him: "Dear comrade you cannot travel. The visa of the military attaché is missing on the open command." By car Karl von Terzaghi searched for the Turkish military attaché colonel F. Bey. After some setbacks he finally met the military attaché in the embassy. He brought the formality to a close within half a minute and reached the railway station again on time. On the platform he met the captains von Kalbacher and Hann (?). "The farewell from both was my farewell from the army. Two officers with heart, intelligence, humor, energy, and zeal. Comrades as one can seldom find."

With the Balkan train via Semlin (Zemun) and Sofia, Karl von Terzaghi arrived at Constantinople after an entertaining ride. Constantinople, present-day Istanbul, in ancient Byzantium, largest city and main harbor of Turkey, is located at the southern estuary of the Bosporus into the Marmara Sea. The city possesses an oriental flair with its many buildings designed in the Byzantine architectural style. Three districts lie on the European side, with the center Stambul with the famous Hagia Sophia and the sultan's residence Topkapi Sarayik, being the most important one. On the Asian side one can find marvellous villas with nice views. Constaninople, settled between the Mediterranean and Black Sea, between Europe and Asia, is a cross-road of old and important traffic and trade routes and radiates a particular atmosphere. Already, on September 12, 1916, Karl visited his future sphere of action, the Ecole des Ingénieurs. The school presented "the systematic order of a Bohemian-pub. The people were as motley mixed as the furniture in the room of the boss ... Two young professors for whom the inclination to scientific research was in no way written on their foreheads, discussed eagerly the kind of questions which should make the candidates waiting in the background happy." Hofrat Professor Forchheimer had a strong position in the engineering school. His right-hand man was a small man with the characteristic face of a clever sales man, named Sadik. In the background,

Karl von Terzaghi in Constantinople (Karl first left in front row)

Karl von Terzaghi around 1916

the predestined successor of Hofrat Forchheimer, a young engineer, waited for Karl von Terzaghi. On September 14, Hofrat Forchheimer accompanied Karl to the Ministry for Foreign Affairs. On this day, Karl signed the contract with the minister of foreign affairs, Prince Abbas Pasha. The section leader welcomed von Terzaghi and Forchheimer with great kindness, served coffee and talked a while with Forchheimer about personal business of the Ecole. Then the section head led them to the minister. Abbas Pasha was sitting with some high officials at a broad writing desk. "An operetta prince whose presence necessarily has to produce amusement." Karl had a very pleasant conversation with the minister. In the course of their dialogue, A. P. asked him to get involved in an irrigation project because of his rich experience in this field. Hofrat Forchheimer commented on the conversation very positively.

On the next day, in the morning, Karl von Terzaghi had an audience with the Austrian ambassador. In the presence of the German ambassador and Hofrat Forchheimer they firstly conversed about more or less unimportant subjects. After one hour Karl von Terzaghi and Hofrat Forchheimer were invited to breakfast. Karl was very pleased with the reception of the ambassador and he enjoyed the atmosphere and the conversation.

After the reception he rode with Forchheimer to the American Robert College in Bebek where they were welcomed by two professors. The buildings and the campus reminded Karl of Portland. The college governed "from its hill the whole wonderful world of the Bosporus. Wide lawn places gives the youth the opportunity to play sports beneath an open sky. A splendid, broad street leads through gardens to Bebek. There is plenty of money, only one thing is missing, imagination and scientific spirit. The boys learn their textbooks, make experiments with their machine tools and the professors do their handicraft in such a way as if they were intellectual eunuchs. No trace of creative activity or also only of a need for such. Nothing more than making money by teaching. Possibly a certain joy at the growth of the number of students and at the school statistics." It was fortunate for Karl von Terzaghi that his diaries were not available at that time so that nobody at the Robert College could read his less than flattering remarks. Otherwise the authorities of the American College would certainly not have offered him a position after World War I.

In March, 1917, Karl von Terzaghi drafted a summary of his last half year: "A half a year of continuous, hard work behind me. The lecture notes for my three courses already written and the plan for the 'foundation' clear in my head. A short pause in one of the most fruitful years of my life, a year of harvest."

He had developed a schedule for investigating fluid-filled *deformable* porous solids. According to his autobiographical notes from 1932 he stated: "that also recognized authorities in the field of civil engineering have no basis for the assessment of the expected settlements of a building. The view of the experts relies only on theoretical arguments without conclusiveness."

Furthermore he remarked in his autobiography: "In the following two years (after he had arrived in Constantinople, the author) I had for the first time the opportunity to leisurely prove, the theories relevant to civil engineering and to compare them with the rich experiences which I had already collected. In the course of these studies, I gradually recognized the reason for the failure of my efforts hitherto in the field of technical geology. The reason lay in the ignorance of the relations which exist between the pressure effects and the corresponding deformations of the soils, and on the lack of test methods whose application yields an immediate and numerical explanation of the technically important properties of the soils. The earth-pressure theories were inappropriate because they only considered the collapse states of the equilibrium of the soils without going into the deformations, which precede the reaching of the collapse state, and the purely geological consideration must fail, because two geologically equivalent layers can show very different strength properties according to their structures. As a result of this finding, the next task was clearly defined. It consisted of the creation of a theory of strength for soils. The task could only be solved via experiments."

In the aforementioned diary entry Karl also described his relationship to his mother and Olga: "Mom wrote to me that she is feeling weak and miserable. Just now, when I am gripped by an aversion to an unnatural devotion of twenty years, the first free, uncharitable, honest letter sent to her. That is fate. What I have written to her up to now was without content. Enumeration of facts and nothing of my innards. Why? Mom was never young with her children and later, everything which appeared great and pure to me was ignored. No art in the house, no free air. No relation to religion, neither warm nor cold. Grumbling at acquaintances, insincerity in consorting. Between the ages of 16 and 17, I already accustomed myself to carefully hiding all that seemed admirable to me , with the intention of not seeing it profaned. No understanding for my purest ambitions. Only the useful, practice! And when it came to marriage, no hope, understanding for the true motives and for noble-mindedness to be found, solely efforts to get along with small-minded prejudices. Now, everything has to come out, no matter what the cost. Mother is the only human being with whom I have been on insincere terms. Now, rigorous action must, finally, be taken also with this lie!

What beautiful times I could have with Olga, if it weren't for the coal problem, food problem and all this stuff of bagatelles. It is as if the marriage developed all pedantic features of my character. All the better. So they become visible and I can begin to fight them, as difficult as this is, because they sit deeply." In the midst of March, 1917, he finished the manuscripts of the French lectures. He stated that he had only few possibilities to include his ideas about technical geology in his lectures. His plan was, through persuading representation of the problem, to attract the interest of a great number of engineers. Furthermore, he tried to become the leader of an expedition which was to investigate a river concerning its navigability.

In August, 1917, he was reminded of his eventful career. In those days he met one of his old teachers who saluted him with the words: " 'Yes, you were always a somewhat adventurously disposed human being and have very often changed. You have studied mechanical engineering, have been a geologist for some time, or a railway builder or now road and railway constructor'. Something like this hurts. This man has no idea what it means to broaden his knowledge autodidactically and to give oneself over to the scientific goal with consequence. I have borne the shrugging shoulders lasting for years with patience nearly a decade and continuously changed and started fresh because I felt my knowledge was not wide-ranging enough ... Nobody has known and nobody wanted to believe that behind the seemingly so planless life persistent and systematic work is hidden, which is directed to a firm goal and only one man felt it without recognizing it: Forchheimer. The others, however, these dilettantes of life who believe the sacrifices which they have made in their careers have been sacrifices to science and who in a lifetime have not brought science a step further and only have vegetated chewing the cud, they still permit themselves disrespect."

Around the end of October, 1917, it seems that Olga was with another man (see Goodman, 1999). Karl reported in his diary about a pleasant, happy day with Olga and a third person on an island. Moreover, he noted suddenly some strange thoughts in his diary: "What charming creatures there are among the girls! If one thinks about this fact one can conclude that jealousy of the physical being of a woman is unworthy of a civilized man. By this jealousy one assigns to her the position of a prostitute, for whom one has rented the exclusive right of use and has observed an infringement on his property through infidelity." In the following paragraphs of his diary he remembered all the charming women he had loved and he listed in detail "the wonderful bodies of women which I have enjoyed ... This gathered feminine beauty exceeds many times over, what the individual woman can offer of her charms, ...

The infidelity of man is no crime, no offense, but a weakness. The man who is proved guilty of an infidelity, feels ashamed as if one has caught him eating on the sly, because of an obvious lack of self-discipline and earnestness. Regarding the woman it would be just so, if Nature hadn't formed both sexes differently. For the woman the corporal love goes namely hand in hand with the ... emotional. At the beginning the shame of the woman, facing her beloved man, is more strongly marked than in her relation to other individuals. The relation reverses, however, after overcoming the first barriers. If a woman nevertheless cheats on her husband, she has also separated innerly; therefore the bad odor which sticks to the violation of trust. The loyalty of the woman to the beloved man is a progressing, unconscious sacrifice which takes place like each act of true virtue with obvious necessity. The woman disclaims all sensations through this sacrifice which the relation to different men could grant. However, the physical love of the woman does not offer the man much more than every other beautiful woman could also offer him. The last reason for his preference for the chosen woman must thus lie in the sensual relations." In this sense he continued with his pseudo-philosophical musings in his diary from 1917. Obviously, Olga's turning to another man must have hurt Karl deeply. He, who always considered women as God's gift to men, realized that he had lost his nearly only passionate and relatively long lasting love. As usual in such situations, if unpleasant occurrences touched his self-confidence, he tried to paraphrase and to mask incidents unfavorable to him and made other persons or circumstances responsible for his misery. His lamenting in this case fills 24 pages in his diary of 1917. At the end of October he took up again his reflections over 7 pages on the relations between men and women and their sexual behavior.

In November, 1917, he gradually calmed down and he finished an article *Kultur oder Zivilisation* (Culture or Civilization) and concentrated on his lectures for the students. Moreover, in January, 1918, he reported that he was deeply involved in scientific work. At the end of December, 1917, he had solved the problem of stress distribution under slits. Thereafter he investigated the cause for the horizontal outlet of landslide in clay and other problems in foundation engineering and soil mechanics. On January 8, 1918, he received a letter from Professor Pöschl with the inquiry whether he would take over the chair of hydraulic engineering at Prague University. Karl von Terzaghi was very happy and on the same evening he wrote back and gave his acceptance ... Moreover he sent an overview of his current scientific work to Professor Pöschl.

He continued his scientific work intensively over the next weeks, trying, for example, to establish the cause of the phenomenon of quicksand from the theoretical, as well as experimental side. On January 21, 1918, he gave a talk in the Mathematical Society on his new theoretical results which was attended by several professors, e.g. Professor Forchheimer, Nielsen and Ludendorff. Forchheimer was impressed by von Terzaghi's new ideas.

After days of great activity, Karl fell back again into a state of lamenting and he expressed his sorrows about Olga's possible separation from him. "If Olga is separated from me, she will lose every mental contact with me and she will have neither the wish nor the ability to share an experience internally with me."

In his diary entry of February 12, 1918, Karl von Terzaghi seemed to mention explicitly Olga's relationship to another man: "She had met him only five or six times, one time outdoors and two times in his apartment. When he wanted to begin addressing her with Du (the intimate you, the author), she did not respond to it and that's final. Before they saw each other every day, in the morning or in the afternoon, however it came, it happened. A strange relationship. Never an intimate talk and he didn't have the courage to demand something that could hurt her or disgust her. The whole thing was a product of unbound love of life and of the wish to challenge the whole

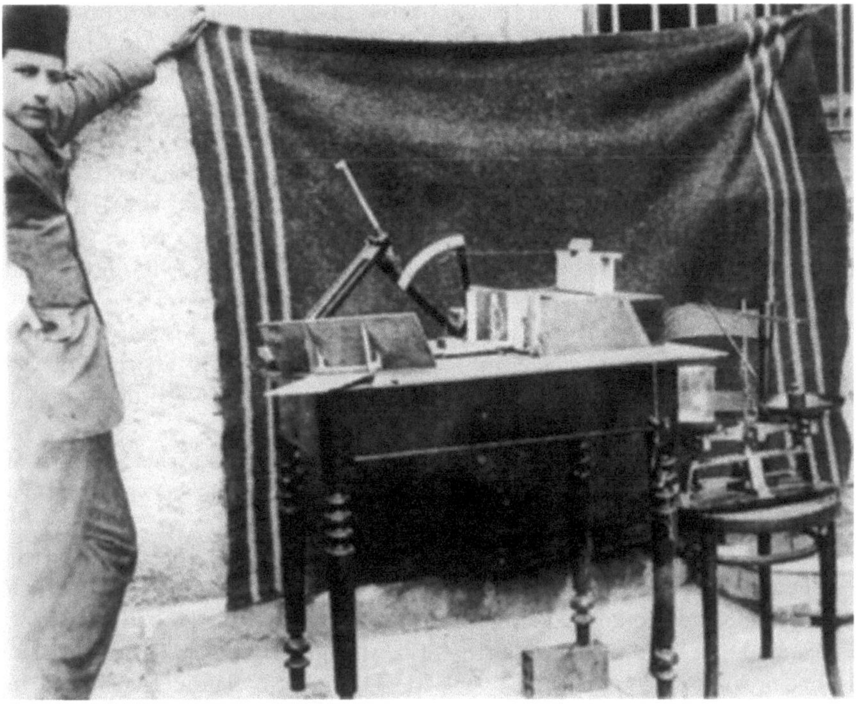

Von Terzaghi's earth pressure test in Constantinople, 1919.

world. He was not weak-willed, though, he seems to have an uncertain feeling of shyness facing her. A tender, fine human being around thirty. He will certainly remember her only with great respect."

During the spring and summer of 1918, Karl von Terzaghi finished the essay *Culture or Civilization* (the complete title was *Nationalist-Racist ("völkisch")-Culture or Civilization of Foreign Races*, and the essay contained obviously racist ideas, the author), which he dedicated to Professor Wittenbauer, who had influenced his writings to a great extent and which he sent to the publisher S. Fischer in Berlin. Furthermore, he began his work on the earth pressure theory in earnest and did some tests on gliding surfaces.

From Austria he received bad news. The returning Bolshevists had devastated the country and the people were suffering from hunger, with only two small loaves of bread each week.

During this time his financial situation seemed to be in very bad shape. His income was low and his private means were exhausted. Therefore, he wrote a letter "to the authorized Excellence" of the k.u.k. military with the request to employ him again for active duty in the Austrian army.

At the end of June, 1918, he formulated the first draft of his new earth pressure theory. He gave a talk on this topic. However, he was not satisfied because he believed that he had not spoken well. His thoughts were obviously still too fresh. He spoke too fast and forgot important points. After the talk was over Hofrat Forchheimer – as well as other colleagues attending the talk – was silent. However, when Karl was alone with Forchheimer, the Hofrat congratulated him and stated: "Your work is fundamental." Back at home Karl enjoyed Olga's playing of Grieg's Peer Gynt suite on the piano.

At the end of June and the beginning of July, 1918, he wrote a short priority announcement of his new findings. The complete new earth pressure theory was finally published in 1919 and 1920 and found some interest in civil engineering circles.

In August, Karl received a letter from the publisher S. Fischer, rejecting his article *Culture or Civilization.* "We find the ideas and attitude of the author valuable; however, the manner of the representation has brought about that the work appears exhausting in its witty sayings."

In the same month he also received an open order from the Ministry of Education that his stay in Constantinople was over and that he was to return to Vienna.

The order came undoubtedly at the wrong time. Karl von Terzaghi felt that he was in the most creative phase of his life, which he would never reach again. He decided to go to Vienna immediately to clarify the bad situation. In Vienna he went from one department to the next, though with no success. He had a lot of free time and spent it visiting Graz again. He arrived in Graz at 2 a. m. in the morning. His mother received him with great love and tenderness. He found her as thin as a skeleton as a result of the bad provisions during the war: only 60 grams of fat per head a week, meat two times a week and the amount of dried vegetables was miserably proportioned. During the next days he met several former friends, fraternity cronies. He recalled his childhood and visited the scenes and haunts of his youth. "Yesterday, in the evening, when I returned, I sat nearly more than one hour in the room in which my grandfather lived at that time, the good spirit of my being, and gave way to old memories; every piece which is in the room, is placed in some relation to my childhood. In this room grandfather raised the best and the holiest which I possess, the absolute strength of character." Moreover, he had time to write seventeen love letters to his wife Olga reflecting on, among other things, his role as husband and reminding her not to be so open-handed in the household. Obviously, he didn't send the letters to Olga. After he had written the last letter he pondered: "Olga is not able to sympathize with what I think and feel. I already knew this years ago and therefore I wanted no child with her; one wishes a child only from a woman of whom one is sure that she is able to be a full companion. Possibly Olga will understand in some years what she still owes me today, in possession of my love. She thinks possibly she is the ruler of the situation, because my thoughts and feelings pass her without a trace. However, this kind of rule must be bought dearly, one pays with a lonesome heart."

At the end of September, Karl von Terzaghi travelled back to Vienna, where he visited several friends. Kalbacher, his long-time friend, told about Karl's companions at Aspern, namely Richard von Mises and Theodor von Kármán. Richard von Mises had formed a strong group with two other members of the staff at Aspern and this group appeared to have become too powerful. "Mises is now pilot in the Southwest." Theodor von Kármán was accused of dealing with external companies and "is sitting today as a technical officer ... in the inner part of Hungary."

Schaffernak gave him some background information about his setback at the Technische Hochschule in Prague and told him that his application had already been reviewed at the very beginning of the nomination procedure with a very negative result. One of the members of the nomination committee had commented on this with the words: "Yes, he should have become a research traveller and not an engineer."

The morning newspapers brought the news that the Bulgarian front had collapsed and they commented on this occurrence in such a free and sorrowful tone that every reader got the impression that the catastrophe lay immediately ahead. In the afternoon Karl visited O. K. Fröhlich, his coworker in several places in Europe, who was very concerned about the situation at the front. Karl was obviously not bothered by the development of the war, since Fröhlich said to him: "This is the most memorable evening we have lived through together and I cannot conceive that you still can deal with your earth pressure theory in these times of world historical importance." After supper, Karl packed his luggage and worked on the final form of his manuscript until the next morning, sent it off and rode then to the north-railway station, rushing for Constantinople. In Belgrade the wagon was uncoupled and the passengers had to stay in the crowded night-wagon. In Bulgaria they rode without lights. The train stopped several times without a perceptible reason. In the East they saw continuous sheet lightening. In Sofia, Karl got a sleeping cabin. When he awoke the next morning after a nearly quiet night, he was already on Turkish ground. With ten hours delay he arrived in Constantinople, where Olga welcomed him. In the evening of October 1, "I visited Forchheimer in his department and informed him that I was waiting for my assignment to Belgrade." Karl had already been told of the possible assignment to Belgrade.

The war came gradually closer to Turkey. Around October 6, it became known that the largest part of the German and Austrian troops fighting in Syria had been captured by the English. The Turks had broken off, on purpose, their contacts with the German neighbor lines and retreated without making their intentions known beforehand. Only four or five artillery officers had rescued themselves from the general debacle.

On October 13, the people in Constantinople saw the historical theater of the arrival of the Entente vessels. Some days later the first French soldiers were welcomed by the inhabitants of Constantinople.

On October 15, 1918, rumors circulated that Adrianople had been occupied by English troops and an Austrian steamer had been torpedoed by a Bulgarian submarine in the Black Sea. Moreover, also in Constantinople the people felt the terror of the war. In a street not far away from the Bazaar, an aircraft bomb exploded, killing around twenty people and injuring fifty or sixty.

On Friday, in the evening, the von Terzaghis had guests and they enjoyed an entertaining evening. After the guests had gone Karl talked with Olga for several hours. He tried to persuade Olga to leave Constantinople for Austria. Olga rejected the proposal, stating that Austria was more dangerous than Constantinople.

Towards the end of the war the press changed its opinion. Day by day one could read articles against Germany. Germany had forced Turkey to take part in the war and now as things went bad for Turkey the Germans were forsaking the confederation – England should be the natural ally of Turkey.

On October 29, Karl von Terzaghi heard the news that the German general Ludendorf had resigned. This occurrence was considered as the first sign of Germany's total surrender. On October 30, 1918, Karl was informed that the armistice conditions were already known: opening of the Dardanelles, evacuation of Syria and Mesopotamia, expulsion of Germans and Austrians. Karl was ordered to join the next transport together with his family. Forchheimer was very dismayed at the news of the coming evacuation.

In the first half of November, 1918, the world had transformed dramatically. Austria and Germany had proclaimed their armistice. "The Austrian army (500 000 soldiers) has been captured and interned by the Italians and the Brenner line and the Pustervalley occupied, according to officer circles; Bolshivism in Germany, report of the corporal E., in Hungary there is anarchy, the higher officers are insulted and mistreated, plundering is the order of the day."

Karl von Terzaghi as a member of a defeated nation was dismissed at the Ingenieur-Hochschule in Constantinople and was without income. Therefore, on December 1, 1918, he tried to get a position at Robert College. His attempt was successful. Robert College, located in the small community of Bebek, north of Constantinople, in former times a missionary school, had expanded its program to engineering and Karl found a position as an instructor in thermodynamics and gas technology.

Monday, December 9, he moved to Robert College and on Wednesday, December 11, he began with his lectures in English. There were only a few students attending the class, who poorly prepared for the material on which he lectured. Some days later he received the preliminary residence permit until January 20, 1919. The residence permit was prolonged with the support of the Swedish ambassador. On December 18, Olga and Vera followed him to a rented house close to Robert College. During the next weeks, Karl von Terzaghi was deeply involved in scientific work. He worked hard on various

Olga von Terzaghi (1916)

problems of saturated soils; for example, he thought about the permeability of finely granulated sand and he supported the view that Darcy's law was no longer valid. Moreover, he designed an apparatus for the determination of deformation diagrams of clay and in the first week of March he nearly finished a paper on dams on permeable foundations as well as the computation of the filtration processes beneath channels.

In the second half of March, 1919, Karl von Terzaghi tried to get settled at Robert College. He put up new paintings in his office and he designed a working desk, closets and an apparatus for the determination of the permeability factor. The students' lack of experience proved a problem. "Students complain about the teaching methods. Years are lost, basic knowledge is missing. Have no idea of mechanics. S. (a colleague, the author) does not know calculation of frames. Nobody seems to have heard of Darcy's law."

Karl von Terzaghi dashed off, furthermore, some notes regarding the political and social situation in some European countries after the end of the terrible war in his diary. "Temporarily, England is making an effort to keep Germany and support it with food, against France and Italy, which want to ruin Germany. Labor leaders in England very reasonable contrary to leaders in the United States. Two months ago huge strikes in Glasgow and other cities; after important concessions to the working-class; high wages compensated by better organization, methods shall be fixed by an industrial board

Olga and Karl von Terzaghi in Constantinople

consisting of employers and employees. In the course of the last ten years remarkable decrease in alcoholism within the working-class. Middle class in England at a deep level, instead of it, the aristocracy stands higher. No arrogance, only exclusivity. Good understanding within the personal relationship between members of the aristocracy and the working-class." Karl von Terzaghi further continued: "the Bulgarian tells, Odessa taken by the Bolshevists. In Hungarian cabinet K. overthrown, Bolshevists fights together with Hungarian Bolshevists against the Romanian, hinder by force of arms the occupation of Transylvania, the train Paris – Constantinople does not run anymore. Rumors, cabinet Clemenceau is overthrown, serious tumult in England and India, fights between Serbians and Croatians. The Swiss reports, that the whole reasonable Swiss working-class has united against the propaganda work of the organs of the German Bolshevist party. The full accord and the effectiveness have been remarkable.

The Entente has temporarily suppressed all political news, the newspapers bring nothing. From the *Observer* one can draw that the moderate Social party (majority of the people) is totally in the hands of the Jews. T. believes nothing can be said about the structure and program of the minority parties, as these change from week to week."

At the beginning of 1919 Karl von Terzaghi installed – with a small grant from Robert College – a laboratory in order to investigate, in particular, the permeability of soils due to change of water pressure.

If one loads statically steel, timber, concrete, stone, and dry sand, and measures the deformations in a small range, one recognizes that the final deformation state is reached immediately. The factor time does not play any role. However, the situation changes completely for clay. This fact had been overlooked for a considerable time. It was Karl von Terzaghi who guessed – inspired by his ability to recognize important physical phenomena – that time must have a decisive influence on the deformation of clay and claylike soils. In a letter to Professor Wittenbauer he stated: "In February (1919, the author) I composed a theoretical work about a weir-construction problem and at the beginning of March, I put together on a sheet of paper what needed to be known regarding the strength and physical properties of clays in order to be able to treat the fundamentals of earthwork on a scientific basis." Without a clear theoretical concept, he would design a series of test devices.

On Monday, March 31, he tested the first apparatus in order to start experiments on sand in rising water and he stated that the K-apparatus (for permeability) and saturation-line apparatus were ready and the bulkhead apparatus was going to be manufactured. On Sunday, April 6, he reported that he had started to make a sketch for a pressure strength apparatus a day before. This should be the best apparatus he had designed. Later this device was named ödometer (derived from the Greek word $\omicron\iota'\delta\eta\mu\alpha$ (for swelling)). He planned to build, in addition, a device for the determination of the influence of the pressure on K (permeability factor), a combination of his permeability device and his ödometer.

The drawback of his devices was that they only permitted one-dimensional load states and that it was not possible to bring three-dimensional stress states into a sample, although such an apparatus was already known in 1911. Obviously, Karl von Terzaghi had not followed the literature on mechanics and experimental research. Thus, he had completely missed the fundamental work of Theodor von Kármán in 1911. This is all the more amazing because Karl knew him already from his service at Aspern. In the course of verifying the failure condition of O. Mohr for brittle materials, von Kármán designed an apparatus which was constituted in such a way that for a cylindrical sample the all-round jacket-pressure and the axial force could be changed and measured independently. This device, manufactured by the firm Friedrich Krupp in Essen, Germany, is now known as the triaxial apparatus. With this apparatus, one gained much more information about the physical behavior of the material to be investigated.

In soil mechanics, the triaxial tests were performed by Ehrenberg, who was in charge of the Department of Foundation Engineering of the *Preußische Versuchsanstalt für Wasserbau und Schiffbau* (Berlin) in 1927. In the fall

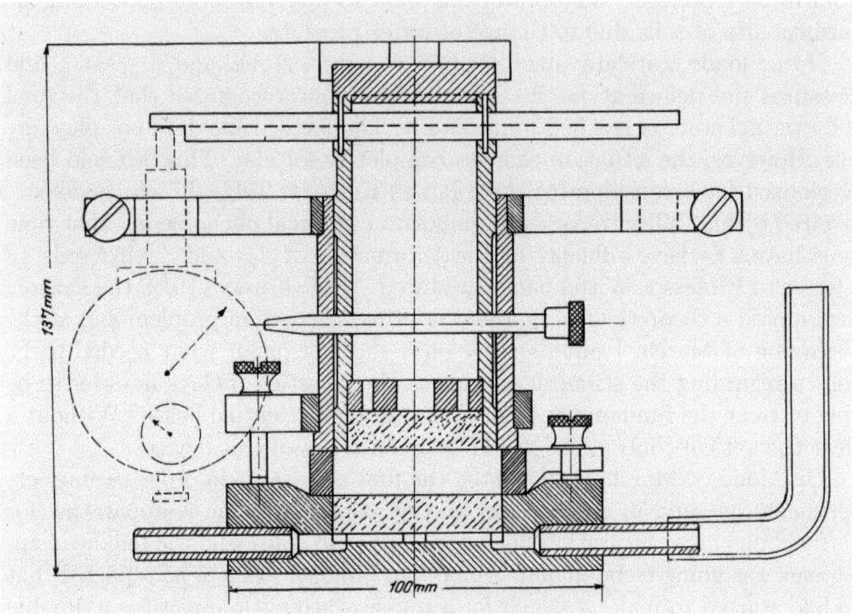

Drawing of an ödometer

Ancient consolidometer (ödometer)

of 1929 Arthur Casagrande visited the *Versuchsanstalt*[29]. He "was greatly impressed by the design of this apparatus." In a report he stated: "Recently an apparatus was designed for the purpose of performing compression and permeability tests on undisturbed clay cylinders. This apparatus is constructed in such a manner that compression tests can be made with any lateral pressure. This possibility will certainly render this apparatus very valuable for research. The compressibility of a soil as a function of the minor principal stress should be known as a basis for a correct determination of the lateral deformation beneath a loaded area. However, no practical apparatus has been available prior to this time." Arthur Casagrande sketched the apparatus in his notes, which included all essential features of the device. Immediately after his return "from Vienna to M.I.T., in March, 1930, the triaxial apparatus was quickly built." In the time to follow a co-worker at M.I.T. changed the design a little whereas Arthur Casagrande, now at Harvard, still used the original apparatus: "For many years these two type of apparatus were simply referred to as the *Harvard-type* and the *M.I.T.-type*." In his paper from the early 1930s Arthur Casagrande gave the impression that he had reinvented the triaxial apparatus which today plays such an important role in geotechnical engineering investigations.

On April 3, 1919, Hofrat Professor Forchheimer said good-bye. The farewell was very emotional; the Hofrat and Karl had travelled down the same road for a long time.

Karl's mother wrote in a letter that his whole luggage which he had sent to Austria via Odessa had been seized in West Ukrainia. In Graz the living situation was very bad: "125 grams of flour per week, 0.5 liters of petroleum per month and since December gas blockade."

Around the midst of April, 1919, Karl von Terzaghi reported about the fate of nearly 200,000 Armenians who were staying in shacks. Gradually they were transported to their home villages in groups of 400 persons per railway. In all municipalities the authorities ordered the Turks and Arabs to hand over Armenian girls and women. Because this was not easy to perform, a premium of one lira per head was announced. The French authorities tried to import Armenian occupation troops. This was the wrong choice. The Armenians, namely, believed they had the right to take revenge against the Turkish people. The French attempt to disarm the Armenians again failed and they fled with the weapons into the mountains. The reputation of the French was badly hurt. The Armenians defended themselves against the Turks for two weeks. After they surrendered, massacre and deportation followed.

During April and May, Karl von Terzaghi performed many experiments on saturated and unsaturated sand and clay. He reported on his results in the *Österreichische Wochenschrift für den öffentlichen Baudienst* (von Terzaghi, 1919) and in the *Engineering News-Report* (von Terzaghi, 1920), in particular on earth pressure and related problems, such as arching. However, his reports

[29] Laboratory

consist mostly of a description of the phenomena, with little mechanics and mathematics. This was his working style also in the next years and thus it was not surprising that his articles made a deep impression in engineering circles because they were relatively easy to read. However, his ambitions were more far reaching.

As can be seen from evaluating his experimental and theoretical investigations he was mainly interested in determining the permeability of deformable soils. Due to his lacks in mathematics and mechanics he was not able to develop a completely new theory of strength for soils. However, his efforts were in a certain sense pioneering although his invention of new terms like *hydrodynamic stresses* calls Goethes statement to mind (Mephistopheles in Faust): *Denn eben wo Begriffe fehlen, da stellt ein Wort zur rechten Zeit sich ein* (For there precisely where ideas fail a word comes opportunely into play).

On Monday, April 28, the last part of his lecture was scheduled. It was very embarassing for him that nobody at the College was interested in the topic he wanted to present. "Such an atmosphere of idiocy, which governs the College, I would hardly have thought possible."

In the middle of June, he was happy about his successfully performed experiments. "In the last week the first coherent test series was finished and gave a rather clear insight into the nature of clay. Against my expectations, the focal point of my investigations has changed from sand to clay and the physics of clay has shifted to the center; it is identical with the physics of water."

In June, 1919, the von Terzaghis moved to another apartment, with a view of the Bosporus. Family life seemed to be in order again. Karl praised Olga's diligence and care and he was pleased how Olga had decorated their new home. He was also delighted by his daughter, Vera, a vivacious and friendly toddler.

In July, 1919, Karl von Terzaghi started his famous experiments on the permeability of clay. "Days of complete disappointment." The difficulties seemed surmountable at the beginning. "Also got to know step by step the sources of failures of the apparatus for the determination of the pressure strength of clay and had to reconstruct the most important parts of the apparatus. Night after night totally exhausted to bed. However, the experience has already shown me that just the absurd test results lead to the most interesting results. *Do you think you are able to find something without searching for it? You will never find anything.* François Vigins!" In his diary entry of August 31, 1919, he remarked: "The last three weeks strenuous activity, my experiments with the bulkhead and the brilliantly successful investigations of the elastic properties of clay, have brought my works to a preliminary completion. The theoretical chapter of my life is about to close and the next years will, I suppose, have to serve to complete my results, make them public and introduce them in technical practice. I considered undertaking an attempt at

Olga and Karl von Terzaghi
at home in Constantinople

first to interest the American government in installing a large laboratory and then calling further ones into existence in other countries."

On October 12, 1919, the von Terzaghi family moved to their new house, because they had to leave their apartment due to the arrival of forty American families who needed the living space. The new house was provided by a Mr. Fisher. The house had a nice porch with a splendid view of Bebek, a small mosque and the Bosporus – and there was marvellous mountain air.

In their new home, Karl received good news from Austria. His long-time friend Schaffernak had gotten the chair for hydraulics I at the Technische Hochschule in Vienna.

At the end of October, 1919, Karl von Terzaghi drew a preliminary summary of the previous years. "The history of the last years of my life shows to me that only then can something completely new be created if the will to completion is stronger than all external circumstances. All that I have created up to today was in principle already available in the fall of 1916, when I prepared my lectures on foundation construction at the airfield of Aspern. Temporarily I feel like a human being again. We have to consider our work from a distance in order to experience joy and to sense the utility."

At the end of 1919, he performed supplementary earth pressure tests with great success. "See always clearer in the earth pressure tests only a means for

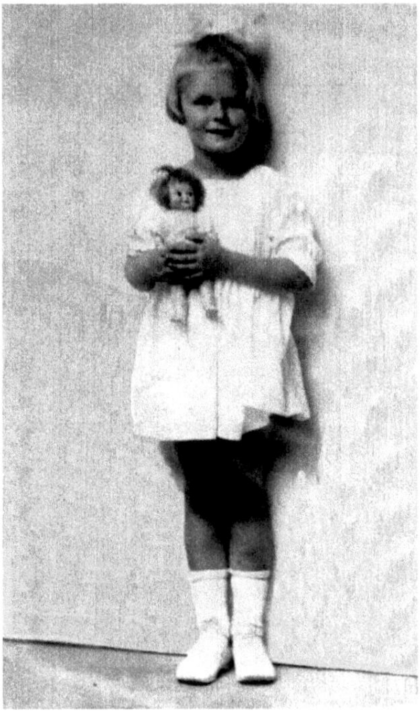

Vera von Terzaghi

the understanding of the processes in the earth." And Karl was happy. This changed unfortunately, at Christmas 1919. Vera caught a flu, their house was in very bad shape and Karl had no notion how the future would be. Moreover, an authority at Robert College told him to extend his lectureship. The administrator informed him that Dean Scipio was of the opinion that Karl was earning too much money and that one could burden him beyond the limit respected up to this time. Mr. Scipio dropped his plan later when he saw Karl von Terzaghi's growing success in his scientific work.

In March, 1920, Karl von Terzaghi was overworked and exhausted caused by his intensive scientific work clarifying the results of his test observations. "Hours of helplessness and discouragement after three weeks of strenuous experimenting. Unaccountable contradictions in the results of the observations about the influence of water smearing on friction number. First at the end of the week I was beginning to see the light in finding that the contradictions could only be explained with the help of my old conviction that the absorbed layers possess shearing strength. Tuesday (March 16) I found the surprising decrease of the permeability number of soft clay with decreasing gradient. However, I was so exhausted that the discovery did not give me much pleasure."

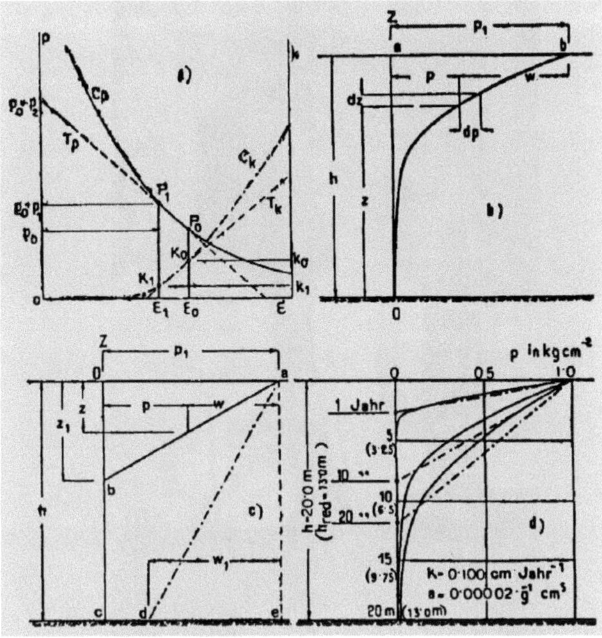

Von Terzaghi's test results of loaded saturated clay, 1923

The week before April 18, 1920, brought great surprises to Karl von Terzaghi. On Monday and Tuesday he tried to formulate constitutive equations for the mechanical behavior of sand. On Wednesday, in the morning, he found the ansatz. He had already supposed a linear relation between "pressure and elasticity modulus" earlier. However, the mechanical behavior of sand was disguised by the influence of hysteresis and elastic after-effects. Now, he quickly saw through cause and effect and he was very pleased at deriving the deformations of concrete from his corresponding formulas for sand. In contrast to his expectations his sand-pressure work became therefore a basic treatise.

On April 29, after a period of hard work and great success, he reflected upon his findings and future. "With the two papers, on earth pressure written in German, I have lifted only a small tip of the veil. It has given me an immense pleasure to see how the first fruits of my considerations, in general, fit in the great frame of the properties of material and how I have succeeded in creating the relation between liquid and solid. Beginning, proceedings and prospective conclusion of the research period of my life can be sketched out in the following way: I have always recognized more and more clearly the last large white spot on the map, namely in the scientific fundamentals of underground engineering and it was for me very fascinating advancing into the unknown regions ... Fortunate circumstances and successful experiments have made it possible for me to lay down the outline of geophysics. This

Karl von Terzaghi in the lab

has made me glad and was worth, in part, the loss of earnings. I leave it to other people to determine the details of the newly discovered fields, even if those others should be awarded the main part of merit. Because I prefer a life according to my manner to the living on of my name after my death. I consider my task as settled as soon as I have lashed up the unhappy representatives of the scholar caste with a pile of publications working further in the direction given by myself. I await this moment with a secrete malicious joy, because I myself shouldn't dig at any price till my death with my nose in the dirt. That should be done by other people. This moment should be reached approximately in one year and then I will exchange the chair with a seating accommodation of my liking. God grant it!" Moreover, Karl von Terzaghi thought over his further career: "I imagine my future either as a consulting engineer or as an underground-engineering-entrepreneur. Three years of continuous occupation with sand and clay do not fail to mark a human being. It develops the instinct for the properties of materials, which is more valuable than formulas."

During the next months he took up his experiments on friction phenomena and filtration again and often got depressed because of contradictionary test results; in addition, he postponed his test with bulk heads until Christmas.

On Saturday, August 21, 1920, Karl von Terzaghi completed his experiments on sand. After weeks of continuous strenuous work he decided to take a break for several weeks with hiking, small excursions and visiting friends and acquaintances.

In October he began to work out the test results of the previous summer and he reflected on his situation: "Now, I think more and more often that the times of scientific research have to come to an end. Either one offers me an appropriate laboratory and assistants of a first class university or I will return to practice ... It seems to me that I have lived only as an individual up to today, caring in no way about social relationships or duties." Gradually his restlessness, which had driven him all these years, returned and he wrote a letter to the American, Allen Hazen, reporting on his work in the previous four years, entertaining hopes of getting an adequate position in the United States. His wish to leave Constantinople was strengthened by the fact that the inner situation in Constantinople was getting worse. "In the town robbery and murder in a public thoroughfare and rapidly increasing corruption of morals. Abuse of children of both sexes, as well as babies, repeatedly reported in the newspapers." Moreover, he was discontented with his wife because of her extravagance. He lamented that Olga was unable to get along with a certain amount of money per month or year. "With the income the demands grow and one remains without savings for life."

In the last quarter of 1920, Karl von Terzaghi finally wrote several papers on retaining walls, earth pressure, friction phenomena and fundamental properties of clays. However, at the same time he also suffered some setbacks in his experimental work – and the response from Mr. Allen Hazen was very negative.

The rest of the year 1920 and the beginning of 1921 were quiet and uneventful. However, on Wednesday, January 12, Karl von Terzaghi was on the way to Brussels, Belgium, doing some consulting work. In the evening, he entered the Orient Express and rode via Sofia, Agram (Zagreb), Trieste, Venice, Milan, Basel to Belgium where he arrived in Brussels on January 17. "The ride from Venice to Milan in the morning of Sunday was like in a dream." After his business was over he used the opportunity to visit his long-time friend O.K. Fröhlich in The Netherlands who, together with an acquaintance, welcomed him cordially. In Amsterdam, Karl visited the Rijksmuseum and was deeply impressed by Rembrandt's *Night-Watch.* Moreover, he did some sight-seeing in Utrecht and Brussels.

Back in Constantinople he tried getting involved in more consulting work. However, he was not very successful. More promising was his attempt at bringing his ideas of "Religion, Science and Life" to the attention of professors, students and other people. On Friday, April 8, 1921, he gave an excellent lecture at Robert College: "Hall filled with listeners. Only a few professors shook my hand and said thanks after the lecture." An American man said: " 'It would had been worth while to come from America after having heard

Karl's lecture ... it is a landmark in the history of the College, new life will start.' "

On May 14, 1921, Olga and Vera probably travelled to Graz or Vienna. Karl felt very lonesome. They both returned first in the middle of October with the Orient Express. Olga looked very relaxed. In the evening of their arrival, Karl and Olga spent several hours in a good mood with an acquainted couple. It seems, however, that the marriage was already broken. There are several signs indicating this possibility. It was already a little strange that Olga and Vera spent such a long time (May – October) away from home and that Karl did not mention the reason and the destination of their trip in his diary. Also Karl's sudden trip to Graz, Vienna and Prague at the end of 1921 and the beginning of 1922 seems to have been an escape from the misery at home. Moreover, on February 15, 1922, Karl told in his diary in a pompous and sentimental manner of a night of love with Olga. Moreover, he displayed his arrogance and self-confidence again as he had done several times before when he was in a similar unfavorable situation: "... and when could you meet a man, who you could love more than me? This, I regard as out of the question." Furthermore, in his diary entry of October 22 he conceded: "Olga. One year has passed since we have known each other as man and wife."

Karl von Terzaghi assumed the aforementioned trip for business reasons, to Vienna and Prague. He combined this trip with a visit to Graz. Karl left Constantinople on December 26, 1921, and rode via Agram (Zagreb) to Austria. In Agram the train was eight hours too late. Here he visited old friends and then continued his ride to Graz, where he arrived on December 29, in the morning at 7 a.m. His mother welcomed him with open arms; however, she was not in good shape. The next days he met several cronies from his fraternity and acquaintances, among them also Professor Wittenbauer. On New Years, 1922, he left Graz and rushed for Vienna, where he arrived in the evening at 10 p.m. In Vienna and Prague he examined some paper and glass factories. In Vienna he had free time and spent it visiting art galleries. Moreover, he enjoyed an entertaining evening with the Forchheimers, his friends for many years.

On March 14, 1922, Karl von Terzaghi received a letter from Miss Obermayer, who informed him that Professor Wittenbauer, his mentor for many years, had died; Karl von Terzaghi was deeply moved. "I see into a world which is new to me; a world, in which one doesn't act, but feels."

The next days were filled with scientific work. He started to bring the manuscripts concerning the test observations with the sand-uplift-apparatus and the theory of barrages into their final form.

The separation from Olga happened on September 27, 1922, when Olga and Vera left Constantinople with the steamer *Adria* "and the connection with Olga is broken off for the present." Karl mentioned this event briefly in his diary on his birthday.

Karl von Terzaghi, who obviously wrote his diary entries of the 1920s later, did not mention any stress in the relationship with Olga in the times around 1921 and 1922 and the remarks about Olga's and Vera's departure appeared briefly and quite unexpectedly. One can only guess that he was deeply hurt by the broken marriage and therefore he was not able to reflect during that time on his misery.

On October 2, 1922, Karl celebrated his 39th birthday, some days after Olga and Vera had left him. In his diary he pointed out: "I must thank the Creator that I pass the threshold of the 40th year of my life as a mature man who has made his talents unfold and has already realized to a large extent the goals, of which he dreamed in his youth. In this summer I had the feeling of being on top of life. My achievements are beginning to receive the recognition and attention which they deserve. The publications of the total results of my previous research and thinking ensured. And the unnatural relationship to my wife cleared up. On September 14, I arrived in Constantinople (He had been in Budapest, the author). There following two weeks appeared to me like one year as a result of the variety of the events. The old love to Olga struggled with the indignation at her behavior and the indignation succeeded." Karl von Terzaghi then reported very briefly about Olga's aforementioned departure and switched over to a completly different subject, namely to the situation in Turkey and neighboring countries. "In the country most serious crisis. ... English and Kemalists stand combat-ready face to face. In Thrazien the Greek army gathers. The Soviets threaten Romania's borders in order to close via Bulgaria the ring around the Black Sea. The whole English Mediteranian Navy in the Marmara Sea and Constantinople under the pressure of the armed English hand. In the College it is a question of to be or not to be. I, however, have already ensured the future by the restless work of the last six years and I cannot be thrown out of my career anymore. Two years ago such a political crisis would have meant a catastrophe for me. For all this I must thank an indulgent Providence in meekness."

However, his last drafts are not very convincing. It seems that they served to overcome the immense pain Olga's and little Vera's departure had caused and a diary entry of October 22, 1922, shows a completely different Karl von Terzaghi. He complained that he had been very lonesome over the last year. "I have thought of you (Olga) daily, this year, of the woman I have loved so much, and of our small child, Verele." He lamented his previous and then-current situation in over eight pages of his diary and expressed several muddled thoughts and strange statements indicating that he was completely out of balance. It must have been at this time that he asked his sister Ella to come to Constantinople in order to overcome his solitude. However, she visited him first in 1925.

In November, 1922, Karl von Terzaghi described a heavy quarrel he had had with Olga on September 22. The cause of this quarrel is not quite clear. Karl had written a letter to his mother and left the unsealed letter on his

Franz Ruchty, Olga, and Vera in the 1930s

desk. Late in the night Olga came into Karl's room and threw the letter onto his bed screaming and crying. In the course of their argument Olga called Karl a barbarian who wanted to throw her out. "The following days were terrible. Now, it was also clear to her that she must travel soon." However, it was not so easy to get seats in the express train. Owing to the bad news from Smyrna the express trains were sold out already for weeks in advance and the ride through Thrazien appeared not to be safe anymore. "In Smyrna the Turks had slaughtered three days long, in the sign of the green flag. The sight of the roads in the burned down Greek quarter, covered with multilated dead bodies must have been horrible." Finally, Karl got, through the procurement of an acquaintance, a cabin on the steamer *Adria*.

The separation was definitive and Olga put an end to the passionate, and sometimes wild, long-lasting love. She was never prepared to resume the relation to Karl von Terzaghi although he asked her, finally in 1925, to go with him to the United States. It seems that Olga was already engaged to the pharmacist Franz Ruchty in Graz at the beginning of the 1920s. In 1926, Karl and Olga divorced and, on September 1, Olga married Franz Ruchty after his first wife had died at the beginning of 1926 (Franz Ruchty had one son with his first wife, though no children with Olga). However, Karl did not forget his great love during his whole life. And he took care of Olga with medicine and other goods in the 1950s, when Olga got very sick. At that time she was living with her daughter Vera and with Dr. Rößner (Vera's husband), in Garsten, Upper Austria. Vera and her husband were Nazis, and had to hide after World War II.

After the painful separation, Karl von Terzaghi moved to another apartment and he began already to write the first chapter of his book which was to be published in 1924/25. Also his duties at Robert College, in particular his lectures to the students, helped him to overcome the moving memories of his beloved Olga and little Vera.

At the very beginning of November, 1922, in the evening Karl saw in Stambul a huge crowd of people with green flags, flares and large drums and pipes, partly marching, partly singing. On the next morning Karl learnt the reason for the demonstration. The last Osmane Mehmed VI had been dethroned as sultan – thus not only his personal situation had changed dramatically, but also the political.

A decisive role in the political upheaval in Turkey was played by Mustafa Kemal Pascha (Atatürk). Mustafa Kemal was born in Saloniki in 1881, at that time part of the Ottoman empire. He got his first name from his parents, and his second one from a teacher later; in 1934 after his great reforms, the surname *Atatürk* was bestowed on him. Mustafa Kemal attended a military school and later became a famous general. Before World War I he was a member of the Young Turk movement and during this terrible war he acted as the leader of a part of the army in Palestine. After the end of World War I Mustafa Kemal took the lead of the nationalist movement. In order to explain the goals of the movement to the Turkish people he toured through all of Anatolia and in 1920 he convoked the National Assembly in Ankara. In 1921/22 he drove away the Greeks from Smyrna, which they had occupied in 1918. After he had dethroned the Sultan, who went into exile in Monte Carlo, Turkey was declared as a Republic in 1923 and Mustafa Kemal became its first president. In the time to follow he carried out decisive reforms, e.g. adoption of the West-European legal system, introduction of the Roman alphabet and surnames, legal assimilation in status of women, and – as a very important point – the separation of state and religion. His agrarian reform, however, failed for the most part. Kemal Atatürk died in Istanbul in 1938.

On New Years Eve 1922/1923, Karl resumed an old tradition he used to do already in his youth: He sat in his apartment drinking a bottle of Champagne and reflecting on the last year and developing plans for the new. "Dense fog over the Bosporus. Some lights are shining through the haze. The New Year has just begun. The sound of the Golden Horn penetrates muffled through the open window of my room – 1923 has begun. Obviously a year of reaching decisions." Then he reflected on his work and the relationship with Olga during 1922. "At first the ideas about the foundations of the weirs were brought into clear form, which I had carried in me for four years. At the end of May the paper was finished. During its writing first conflict with Olga – her indifference considering my hopes and efforts led up to it. It strengthened quickly and in the course of barely two weeks had led to a breaking-off." There remain some doubts concerning Karl von Terzaghi's statement about the moment of the dissolution of his relationship to Olga. It is obvious that the break-up had already happened earlier.

A decisive step in his scientific efforts was taken in 1923. In this year, Karl von Terzaghi (1923) published his famous paper on the permeability of clay. This was the first step out of four remarkable achievements Karl von Terzaghi reached in the years 1923 to 1925 based upon his experimental work and his keen observations of physical phenomena in sand and clay. The second one was his talk at the congress for mechanics in Delft, The Netherlands, in 1924; the third and fourth steps were performed by him in 1924/25, when he published his first book on soil mechanics and a series of articles on the same subject in America.

In his first step, he developed a differential equation which governed the approximate description of the spatial and time-dependent change of the porewater pressure. The derivation of this equation took a long time of intensive thinking and was indeed not easy: "In 1923 my work came to a standstill again. It was a question of the mathematical consideration of the process of gradual compression, which clay suffers under the influence of a constant pressure. I had grimly kept at this task to such an extent that I failed in my duties for about one month and I tried to solve the problem nearly every day until late into the night. The attempt failed and I decided to publish the results of my investigations up to then with the intention of leaving the solution of this central problem of soil mechanics to a successor with more luck.

Half a year later, while I was busy with the record of the planned paper, I succeeded in solving the task without trouble after half an hour of deep thought. It consisted of the differential equation of the consolidation process, which today is the basis for all investigations concerning the gradual settlements of buildings on clay. First, after the establishment of this equation, the attempt to reveal the strength properties of soils could be denoted as successful."

Von Terzaghi reported on the derivation of the partial differential equation of the consolidation process to the Academy of Sciences in Vienna on June 23, 1923. The derivation is contained in von Terzaghi's (1923) paper *Die Berechnung der Durchlässigkeitsziffer des Tones aus dem Verlauf der hydrodynamischen Spannungserscheinungen*. In this work, he was not particularly interested in settlement problems but – as is indicated by the title – in the calculation of the permeability coefficient of clay. The transfer of his ideas to settlement problems was done by von Terzaghi and Fröhlich (1936). At first, he stated that the permeability coefficient of clay was dependent on the water content, and the water content dependent on the pressure in the clay. Furthermore, he showed by the direct experimental proof method, the validity of Darcy's law for very small pores and lesser permeability. The direct proof method failed, however, for permeability coefficients $k \leq 0.06\,[cm/year]$. For this reason, he looked for a new proof method: "Such a method was found in the observation of the time dependent course of the hydrodynamic stress phenomena in clay."

Von Terzaghi explained the term hydrodynamic stress phenomena as follows: "By hydrodynamic stress phenomena, the author means the delays which the stresses caused in the clay by an external force suffer by the resistances against the outflow of the squeezed-out porewater. The local variation of the stresses in the clay causes a change in the water content at the same place. Since the change of the water content requires a flow out of the water, and the flow of the water according to the very small permeability requires a considerable gradient, the effect of a pressure variation is firstly shown in the occurrence of local differences in the hydrostatic pressure in the porewater. These pressure differences disappear first in the course of time. The time-dependent process of the stress balance is caused by the stage of permeability. The permeability coefficient can therefore be determined from the time dependent course of the balance, if all other important factors are known."

Von Terzaghi's derivation of the partial differential equation (1923) is, in parts, obscure. Obviously influenced by the differential equation for the description of heat propagation (Fourier's law), which has the same structure mathematically, he succeeded in developing his partial differential equation more through intuition than through ensured mechanical principles and mathematical rules. Furthermore, in his basic paper, some terms are introduced in a very unscientific way, being in parts not at all defined, or first explained after having already been used. Thus, one can hardly follow his paper in some sections. Even so, the fact that the water-saturated clay body was a mixture was clearly known to him; he stated later: "As a matter of fact, analysis of the phenomenon leads to new and very important conclusions. In order to perform it, we have merely to keep in mind that the elementary laws of mechanics which apply to solids and liquids in general are also valid for the constituents of a mixture of clay and water" (von Terzaghi, 1925b).

However, in 1923 as well as later, von Terzaghi made no use of this finding. He was obviously not well-trained enough in mechanics and mathematics to be able to treat this difficult problem as a two-phase-system. The sometimes-sloppy work (von Terzaghi, 1923) is obviously founded in von Terzaghi's personality: he did not like to work out the final drafts of strenuous subjects. He was, however, an excellent and highly-gifted engineer, with great intuitive abilities, who was able to recognize difficult physical-technical problems very clearly, and could analyze and solve them. This ability was also revealed in his treatment of other problems, such as the solution of the uplift problem. Nevertheless, after he had solved a problem, he showed very little interest in the formulation of the scientific problem and the solution gained, as he declared in the questioning at the Technische Hochschule in 1936 and 1937.

Erdbaumechanik

auf bodenphysikalischer Grundlage

Von

Dr. Ing. Karl Terzaghi

Zivilingenieur und Professor am amerikanischen Robert College in Konstantinopel

Mit 65 Textabbildungen

Leipzig und Wien

Franz Deuticke

1925

Von Terzaghi's first book on soil mechanics

The entries of the diary in 1923 were rather meager. One can guess that the separation caused him more trouble and unrest than he admitted. Only on New Years Eve did he reflect on the year 1923 and recalled some events. "1923 has come to a close – the fourtieth change of the year since I could differentiate light and shadow at the Hradschin in Prague for the first time. A year of slow progressing, though of, obstinate and systematic work. No decisive year concerning events, but a year of decisive preparations. No questions – only answers. Good fortune ? – I have come to know it this year. Steady, positive action. Making clarity of confused facts ... I have not seen Olga this year. The relationship is clear ... Unfortunately, I have spent the summer with strenuous work. Nevertheless: it contained so many events. The fare-well from Constantinople in June. I could barely detach myself from Constantinople. The splendid summer nights, the crystal clear water, the divine calm in Bebek." After these rather sentimental reflections he reported on his stay in Schladming where he was on vacation – and where he had a passionate affair with the "small wild Hilde."

Concerning his work Karl could not state many positive things: "Yes, I have neglected my duties this year ... I can justify this behavior only through the expectation that the results of my work will open a sphere, in which I can more charitably act than in my current one. The year 1924 should bring the change."

During the next months, Karl von Terzaghi worked on his book *Erdbaumechanik auf bodenphysikalischer Grundlage* and he finished it in the spring of 1924. This book was later recognized worldwide as the first essential book on soil mechanics. It contained investigations on the properties of soils, friction between soil parts, strength of soils, hydrodynamical stress phenomena, permeability of sand and clay, as well as statics of soils and on foundations.

The sometimes sloppy work, written obviously in great haste, contained a lot of unclear terms and missprints. Karl von Terzaghi himself was aware of the disadvantages of his first work. In 1932 he stated: "This book features all advantages and lacks of a first work. It is of a piece and is written with the ecstasy which was set off by intellectual penetration of the subject, successful outcome recently. However, it greatly overshoots the mark in many points and lets the properties of soils appear easier than they really are. For this reason I also resisted later the translation of the book into a foreign language, a wish which was expressed by different sides."

In April of 1924, he attended the *First International Congress for Applied Mechanics* in Delft, The Netherlands. All the famous mechanicians were present: Theodor von Kármán (aerodynamics) who knew Karl von Terzaghi from the airfield at Aspern, Ludwig Prandtl, famous for his work in hydrodynamics and aerodynamics, Hans Reißner (shell theory), and the English mechanicians Griffith, Haigh and Southwell among others. Karl von Terza-

Dr. techn. Karl von Terzagi in Delft, April 24, 1924

ghi was obviously nominated by Professor Forchheimer, the excellent expert in hydraulics, because it was not easy to receive an invitation to such an important conference. Karl gave a talk: *Die Theorie der hydrodynamischen Spannungserscheinungen und ihr erdbautechnisches Anwendungsgebiet* (The theory of hydrodynamic stress phenomena and its earthwork technical field of application). How the audience received his talk cannot be answered. Karl von Terzaghi insisted in his diary that the famous German scientist Prandtl had praised him for the extraordinary simplicity and brilliance of his fundamental ideas, but there is no contribution of discussion published in the proceedings of this important conference, as was done for many other talks.

He was back in Constantinople on May 10, 1924. During the next day he had to carry out his correspondence.

In the middle of August, 1924, he was again in Vienna, where he visited many acquaintances and friends, art exhibitions, the theater and other points of interest, like the Belvedere-garden, Schwarzenberg Palais, Stefans cathedral and Karls church. He enjoyed dining with Forchheimer and discussing all manner of topics with this lively and ingenious scientist. He left Vienna with a Danube steamer on August 30, 1924, and headed for Bucharest via Bratislava, Budapest and Belgrade. It was a very smooth ride and Karl found the leisure to read Alfred Wegener and Oswald Spengler, both of whom he admired very much, though in particular, Spengler. He stated that he had mentioned similar ideas as Spengler in his essay *Culture or Civilization* years ago. In Bucharest he left the steamer and made a side-trip to Sinaia, into the mountains, into the province, by train, after he had spent a nice day in Bucharest visiting the geological museum and dining with an acquaintance. In Sinaia, a resort area, the summer residence of the royal family was located. He arrived in Sinaia on September 3, 1924. Romania, like other countries on the Balkan, was very poor. During a walk up to the mountains he saw a village down in the valley: "it is a settlement like those in the Wild West. Shops and plain, but dirty houses, ragged people." After some days of relaxing, reading and hiking, he returned to Bucharest. The next station was Constanţa on the Black Sea.

Back in Constantinople, Karl von Terzaghi had to do a lot of routine work and was involved in consulting. Moreover, he was intensively working on cohesion, permeability and friction problems and writing scientific papers. In the midst of October, 1924, Karl received a letter from Olga, which brought him completely out of balance. "This greediness and selfishness are inexplicable. Almost pathological. 'Scorned' the money that I earn and extorted every penny of that which she has obtained by fraud. My blood boils, if I think of this clever, cool imposture. Will the hour of reckoning finally toll? And under what circumstances? The roads of fate are truely uncharted." He repeated his heavy attacks in a diary entry of December 22 and shortly after Christmas he thought over his coming position towards Olga; he came to the conclusion that the settlement in the United States seemed to be unavoidable. How far Karl was right cannot be proved; there is no corresponding statement available from Olga.

In November, 1924, Karl was again scientifically active and wrote an article entitled *Bausande* (sands for building purposes). Moreover, he planned to publish a textbook about mechanics and he dealt with capillarity. The impetus to think about this important phenomenon came from a letter of a young scientist from the Technische Hochschule of Vienna, J. Kozeny, who had obviously asked Karl about his opinion about a theory he had just developed (see Kozeny, 1924). Karl von Terzaghi responded to Kozeny's letter in a treatise of sixteen pages. The judgement in his diary was devastating. "The man has no sense for proportions. Possibly mathematically virtuoo, but no physicist." The judgement is hardly understandable and not acceptable. Kozeny's work from 1924 is a fundamental step towards a consistent capillarity theory of porous bodies. Karl von Terzaghi's letter seems to have been milder in reality, because Kozeny responded in a friendly tone. On the factual side he stood firmly, however, and rejected the petty objections.

Kozeny (1924, 1927), representive of hydraulics and a member of the faculty of the Technische Hochschule in Vienna, treated the dynamics of the capillary motion whereas von Terzaghi was concerned with the static side. It is really amazing that von Terzaghi failed to mention the excellent two papers by Kozeny, more so since he knew the paper of 1924 very well.

On New Years Eve Karl von Terzaghi reflected, as usual, on the last year: "My work, the fruits of many years of hard work, (presented in his new book *Erdbaumechanik*, which was published in 1924 – see von Terzaghi, 1925 a – the author), was accepted with chilly silence. Rehbock, Pöschl, Müller-Breslau, Kármán (well-known professors at that time, the author) did not even thank me for it. Schaffernack does not mention it with one line, Schmidt (Schmitt?) in New York does not write any more ... Human beings are struck with blindness and worse than that. The people whom I helped to an existence and who found in the moment of need support in me ... have forgotten me at the turn of the year. Olga is obstinate and more selfish than ever. Nevertheless! I can not become bitter. After winter spring comes into the country and he who does not give pleasure to the beautiful properties of the next one, cuts himself in his own flesh. Love, admire and work – what worries you for the rest! Be a human being and as good as your nature allows. With that you help yourself and the others." Karl von Terzaghi's depressive mood continued in the first weeks of the new year. During this time, he visited some acquaintances, made several excursions in Constantinople as well as to the countryside and he enjoyed some silent movies. Then, at the end of January, he received a positive review. After many days of deepest discouragement, he felt newly revived. However, the silence of other scientists, to whom he had sent his book bothered him; and in the middle of March, 1925, he complained: "The *Erdbaumechanik* is obviously a failure. Complete indifference in Europe, rejection in the US, no hope for acknowledgement anymore. Fought with deep grief for weeks, then I pulled myself together. The matter is good and its purpose of answering my questions, was reached. Will not give a damn for the rest. Set about bringing the technical geology into a system, unconcerned whether one recognizes my efforts or not. The time will come, however, I will not experience it anymore ... I am as poor as a church-mouse. However, the zest for work is again back and this is the main point."

In this situation Karl von Terzaghi felt very lonesome and he asked his sister Ella again to visit him; he expected her for the middle of April. After all the misery in the last weeks he, finally, received good news, namely the very positive review of his *Erdbaumechanik* in the Engineering News-Record and he uncorked his last Moët-Chandon.

Karl von Terzaghi's depressed mood caused by lack of recognition of his work and the uncertainty of his position at Robert College would soon dissipate. In the meantime, the dean of engineering (Robert College) namely, Lyn R. Scipio, who graduated from Purdue University in 1908, had written a letter to Dean A. Potter from Purdue describing a young researcher in the department of Civil Engineering by the name of Karl von Terzaghi, who was advancing some new ideas and philosophies concerning foundation engineering. Dean Scipio asked of offering him a position for research in soil mechanics and foundation engineering with a salary of one hundred dollars per month. Also Professor Paul H. Dike of the University of Vermont, former professor at Robert College, supported Karl and wrote a strong letter of recommendation to Dean Potter. Dean Potter conferred with the head of the Civil Engineering Department of Purdue University about engaging Karl von Terzaghi. However, the Department was not interested. Thereupon the dean transmitted the information to Dr. Stratton, then the President of M.I.T. with the suggestion that M.I.T. might be interested in the young researcher. Dr. Stratton was highly interested and he thought compensation of one hundred dollars per month was too modest and should be increased to two hundred and fifty per month. Then, on May 29, 1925, Professor Charles M. Spofford, Head of the Department of Civil and Sanitary Engineering at M.I.T., wrote a very kind letter with the inquiry "as to whether you would care to become associated with our teaching staff on the basis of an engagement for, say, one year as a special Lecturer and Research Associate."

In the evening of April 16, 1925, Karl picked up his sister Ella at the railway station. During the next days he was happy having Ella with him and together they visited several acquaintances, and one woman invited Ella for a one-day excursion to the countryside. In his free time, Karl took his sister to many points of interest, e.g. to the Bazar and an aqueduct. On April 29, he accompanied Ella to her steamer; it was an emotional farewell.

After the publication of his book *Erdbaumechanik*, he worked restlessly on a series of articles in *Engineering News-Record* in which he represented a simplified version of the essence of his findings. These articles, masterly edited by the associate editor F. E. Schmitt, and published in November and

December, 1925, made him well-known in civil engineering circles; but they were also severely criticized.

In this pioneering work on the *Principles of Soil Mechanics*, Karl von Terzaghi mainly outlined the mechanical behavior of clay, which he had studied through his carefully performed experiments at the Robert College in Constantinople. He had laid down his observations and results in eight reports beginning with the cohesion phenomenon, compressive strength, and the determination of the permeability (Darcy's law and improvements) of clay. In his fourth article, he described the consolidation and permeability problems and in the next article he investigated the physical differences between sand and clay. In the sixth article, he presented the stress-strain diagram of sand and a table in which he showed the mechanical differences of sand and clay. In the conclusion of the his fifth report, Karl von Terzaghi stated: "The results of our comparison between the physical properties of sand and of clay are presented in the table. The facts stated in the first and last column of table three have long been known, but knowledge of the causative connection, however, between these two groups of facts is new. It is a much simpler connection than might have been expected. This table, the curves of Fig. 4 (stress-strain diagram for sand) and the partial differential equation (see von Terzaghi, 1923, the author) represent the fundamental principles of mechanics, in a nutshell. They are the key to the physical explanation of whatever properties a soil may display, in the laboratory or in the field."

The seventh treatise was concerned with friction in sand and in clay and he stated finally: "In clay the friction coefficient for medium and high pressures is remarkably constant. For low pressures, however, the value of the coefficient increases with decreasing pressure, because initial friction plays an important part, amounting to about 20 g/cm^2. Rapid change of pressure produces in the liquid component of the clay a positive or negative hydrostatic pressure. The coefficient of friction does not assume its normal value (coefficient of static friction) until the hydrostatic pressure has become zero throughout the whole mass. In the preceding stage of the process, the coefficient of friction (coefficient of hydrodynamic friction) may have any positive value, and is a function of the time."

In his final article, Karl von Terzaghi gave an outlook on the future development of the field and potential problems – and he made a surprising statement: "For practical application, soil mechanics must be applied in the first instance to the following objectives:

(A) A theory of models must be developed. Without such a theory no valid conclusion can be drawn from the results of loading tests on soils or from model tests (as on dams), with respect to full-sized structures."

Statements (B) and (C) were concerned with the classification of soil and with adequate design data.

The eight articles on principles of soil mechanics were clearly arranged and due to his gift for recognizing essential physical phenomena and his ability to keenly observe geomechanical problems, he was able to describe the mechanical properties of clay intensively giving a clear picture of its behavior. In his treatise he avoided the mechanical foundation of the observed phenomena but described them generously with only some easy formulas and figures. Thus, it is not surprising that his articles raised a considerable amount of attention in the circle of consultant engineers and contractors in civil engineering, and he received a remarkable amount of positive responses. However, he was also sharply criticized, e.g. in letters of the physicist E. B. Wilson who had studied some articles and also Karl von Terzaghi's talk at the mechanics congress in Delft in 1924. He found von Terzaghi's contributions to be unsound and of little value. Also T. Merriman expressed harsh objections to some statements in the articles on the principle of soils. Indeed, the flaw in von Terzaghis treatises is that he did not base his findings on mechanics although he treated mechanical problems, namely the motion of material points. For example, it was clearly known to him that the clay body was a mixture (see the above statements; von Terzaghi, 1925b).

This realization was his only contribution to the theory of porous media, although the mechanical mixture theory and the volume fraction concept[30] were already known. His scientific work in this aspect reminds one of such researchers who describe, e. g., the elastic line of a loaded beam by a mathematical function without any mechanical founding. This method may be sometimes alright for a special example. However, it is of no value for a general theory and thus, Karl von Terzaghi violated in a flagrant manner his objective (A) mentioned above.

The next weeks were occupied with his duties in the Robert College. Gradually he got more and more nervous. At the beginning of June he realized that the College would be closed and he did not know where he would be able to spend the summer and the next Christmas. Finally, on June 29, 1925, he received a telegram from the Head of Department of the Massachusetts Institute of Technology (M.I.T.), Dr. Spofford: "Twenty-five hundred dollars for the academic year. Considerable free time." For a moment this left him speechless. "This was the decision. A paltry offer. Nevertheless, the road into the open, after eight years of quiet seclusion." Karl von Terzaghi immediately undertook great activities: he rode to the embassy for a visa, to a bank, and he responded with a telegram: "Accepted ..." and he added in his diary "With this my fate was decided."

After having sketched Karl von Terzaghi's strenuous and sometimes exhausting work in Constantinople with the aim of creating an earthwork mechanics in the years from 1916 up to 1924 the time seems ripe to evaluate his achievements. Although he very often spoke of earthwork engineering, he essentially wanted to establish a new science in what he at first called earthwork

[30] The volume fraction concept relates the volume elements of the individual constituents of a saturated porous solid to the volume element of the control space which is shaped by the porous solid.

mechanics (Erdbaumechanik) and later soil mechanics (Bodenmechanik) (see the later mentioned speech to the Boston Society of Civil Engineers). The later term makes it clear what he intended, namely not only to describe the structure but also to investigate the mechanical behavior of soil including filtration problems, settlements and other physical phenomena. In order to measure his achievements in this respect, some thoughts about science and mechanics will be outlined.

Science, in its original sense in the philosophy of antiquity and the Middle Ages, was assumed to put together systematically all previously gained knowledge (objective truth) as it is known from the arts, the humanities, medicine and in parts from engineering-sciences. The decomposition of this universal concept of science commenced with the development of the modern natural sciences in favour of a stronger emphasis on the individual sciences. At the same time the mathematical scientific method became the model for all scientific methods with the aim of predicting theoretically observed phenomena. Scientific method means the unbiased, methodical, value-free, understandable, duplicatable verification of each statement, the possibility of criticism – and purity, i.e. not being directed to any specific purpose. The main requirements in this list are without doubt, methodical proceeding, verification of each statement and the development of science not for special purposes but in order to seek the truth. In particular, the last point, posited by the Greek philosophers, has brought great success for occidental culture. Due to research without referring to a special purpose, the development of geometry in Ancient Greece was made possible. This decisive progress in mathematics was never reached, for example, in Egypt, although the Ancient Egyptians, forced by the annual flooding of the countryside of the Nile, had installed and developed an effective surveying practice. However, they adhered to a geometry which was only valid for one purpose, namely to survey the country again after the flood.

Classical mechanics was created and developed on the basis of the primary requirements stated above, according to a scheme by the author, already partly introduced by Galileo in the seventeenth century, which is very similar to the four-stages of modern scientific theory (see Popper, 1994):

1. Experience, observation, historical studies and recognition of problems;
2. Attempts at formation of theories;
3. Attempts to eliminate false theories through critical discussion including experimental and numerical verifications;
4. New problems which arise from the critical discussions.

In order to create a new science in a field of natural science, it is obvious that the problem first has to be recognized. This is only possible if one has great experience in the field in consideration and the gift of keen observation. In particular, one has to be familiar with all ensured findings in this field and in many cases also with their historical background, which can be very helpful. Then, hypotheses and theories have to be created. They have nothing to do with practice rather, they are in the modern sense, findings gained by reasoning, explanations of correlations and facts, received from hypotheses on which they are based, which themselves have to be verified and systematized. The theories have to be carefully and critically discussed; in particular, they have to obey axioms (objective true laws) and have to be experimentally proven – if this is necessary. Moreover, numerical investigations have to be performed in order to prove that solutions of problems of the new theory are possible. This method has been mainly developed over the last decades.

From the discussions it may happen that new problems arise which have to be treated in the aforementioned manner.

On the described basis of science, we can rate Karl von Terzaghi's achievements up to 1925. Before doing this his results should be summarized. This had been done by von Terzaghi himself in a lecture given to the members of the Boston Society of Civil Engineers on November 18, 1925. "The problems of foundations are as old as any with which the structural engineer has to deal. However, during the last one hundred years structural engineering became an elaborate and very intricate science, while in the vast field of earthwork engineering nothing has been produced except a set of pile-driving formulae and an academic theory of earth pressure against retaining walls. Symbolically speaking, these theories fit into the frame of modern civil engineering as Trinity Church fits into Broadway. The structure is very venerable, but its foundations are just strong enough to support a one-storey building. Hence, before erecting a more elaborate structure, one has to dig deeply below the old foundations and to provide a more substantial substructure.

This has been done by careful study of all those physical factors which cause the difference between the soils the engineer has to deal with and the ideal material invented by the authors of our classical earth-pressure theories. These factors can be assembled in the three groups: first, the forces acting at the points of contact between the soil grains; second, the strain effect produced in soils by external forces; and third, the effect of time on the development of strain. ... – All these physical factors are intimately connected with the structure of the soil." In the following paragraphs von Terzaghi described, at first the structure of cohesionless sands and he remarked that it ranges between two extreme limits – the dense granular and the loose granular structure. Moreover, he discussed the case where the structure is instable and the possibilities of compacting loose sands and of loosening dense sands. A special section is devoted to the structure of very fine-grained materials and the mechanics of grains affected by pore-water, in particular, also in the capillary range.

Karl von Terzaghi described the structures and some mechanical effects with very clear words as he had already done for other observations in nature.

In the main part of his talk in Boston, he turned to the mechanical phenomena of loaded soil bodies with empty and water-filled pores. He commenced with the statement: "A study of the strain effect produced by external forces in soils has led to the result that the relations between stress and strain in soils are almost as simple as they are for solid granular bodies. The only difference which was found to exist is this: for solids, the modulus of elasticity and the resistance against shear is a constant, merely depending on the nature of the material. For powders both modulus of elasticity and resistance against shear increase in direct proportion with the intensity of the internal pressure ...

This simple fact covers the vast field of stress-strain relations in soils, provided the voids of the soil are filled with air.

... – However, if the voids are filled with water we are obliged to consider a factor absent both in the mechanics of soils and of dry powders. Suppose the strain effect consists in compression. Compression means a decrease in volume. Both water and grains are practically incompressible, hence the compression involves a flow of water through the voids from the interior of the sediment towards the surface. No flow of water through porous bodies is possible without a hydraulic gradient. Hence the external pressure produces first a hydraulic excess pressure in the voids' water, and the speed with which the strain effect takes place is determined by the speed with which the water can escape towards the free surface. In coarse-grained powders, the resistance against flow is very small, and as a consequence the strain effect produced by external forces occurs almost as rapidly as it does in solids. However, the smaller the grains the greater is the resistance the percolating water has to overcome, and for colloidal soils this resistance produces a lag of compression of months or of years, depending on the degree of permeability and the thickness of the deposit. A theoretical study of this phenomenon has furnished the following result." Karl von Terzaghi repeated his partial differential equation which he had developed in 1923 and he remarked: "The relation which exists between the time and the hydrostatic excess pressure ..., acting in the capillary water at the point P located in a depth z below the surface, was found to be determined by the equation – ..." Here the differential equation followed. Furthermore, he made some comments concerning the analogy of his equation with the corresponding differential equation of the theory of linear non-stationary flow of heat through homogenous bodies and described the onset of water flow in a loaded saturated porous body. In particular, he studied the consolidation of mud deposits and the observations during the driving of piles. Moreover, he made some remarks about the permeability in various soils he had studied in the preceding years, and extensively described foundations of weirs in permeable ground.

He closed his talk with a summary of his previous work and gave hints for doing further research and education. "Gentlemen, at the outset of my lecture I compared the classical earth-pressure theories with Trinity Church. The urgent need for a more substantial building is unquestionable. What I have done has been to undercut the old foundations and to brace them against more resistant substrata. In so doing I was alone, with limited funds and without assistance. This fact may account for the rudimentary character of what I could present. The results are promising, but the major part of the work remains to be done. Hence I wish to conclude my lecture with a few remarks concerning what should be done next for developing the young science.

For advancing a special branch of physics or chemistry, it is sufficient to provide funds, fellowships and to establish new laboratories. There are ample men who can do the work, because training in physics has been standardized. However, in soil mechanics the situation is far more complex. For producing valuable results in this field of applied science, a man must be at home in at least five or six different sciences, some of them far remote from the field of civil engineering, and in addition he must have a broad, practical experience or else he cannot discriminate between what is essential and what non-essential. The men who nowadays fulfill these requirements are exceedingly scarce, because their expertness has not been produced by systematic training but by patient self-education, guided by instinct rather than by a defined program. Without such men, nothing worth while can be accomplished. We learned this fact from the cumulative results of related efforts made during the last ten years. Ten years ago we were more optimistic concerning this point. Furthermore, in physics or chemistry objectionable results if published do not harm, because they are pretty soon disproved. In earthwork engineering such publications are utterly detrimental, because very few engineers are equipped with the knowledge required for competent criticism.

Hence, what we need more than anything else are measures of an educational character. Young men who desire to engage in underground engineering should get the proper guidance and an opportunity to acquire all the knowledge required for competent work in their future field. Competent work in this field requires among others far more thorough training in applied geology, physics and physical chemistry than the structural engineer needs. The students who get this training would represent a generation of engineers better fit for shedding light over the obscure field of earthwork engineering than is ours. An *individual* may furnish the ideas, but the bulk of the work must be done by a properly educated *generation.*"

Concerning Karl von Terzaghi's criticism of the old theory of earth pressure against retaining walls and earthwork engineering, in general one can state that his cutting remarks were not justified. His historical remarks about the achievements of these authors stem to a great extent from the work of Kötter (1893). Karl had obviously neither read the original works of Coulomb and Rankine nor was he familiar with the fundamental investigations of Woltman and Mohr (for further information, see de Boer, 2000). Coulomb and other authors considered the earth-pressure-problem as a minimum/maximum problem based on a simple failure condition which fits test results very well. Today, those minimum/maximum problems are well-known in limit design theories (within plasticity theory) as the *kinematic method*, which yields approximately the earth pressure, unfortunately on the unsafe side. With the *static method*, one gains an approximate earth pressure on the safe side. Thus, with the kinematic and the static methods one attains two approximate solutions which restrict the real value of the earth-pressure. Also Professor Krey (1927), director of the *Versuchsanstalt für Wasserbau und Schiffbau* in Berlin, remarked in his article: "If one tries today to construct a contradiction between the old classical earthwork statics and the new soil physics, so it is a misuse of the word earthwork mechanics and a complete misunderstanding of Terzaghi's work. Both support each other and they are the same at the bottom. The soil physics forms the fundamentals of the earthwork statics; and the earthwork mechanics striving for a deepening of the old correct knowledge, is, however, not in contradiction to it."

Today, the approximate character of Coulomb's and Woltman's solution of the earth pressure is well-known and is still used.

Nevertheless, von Terzaghi recognized that the deformation state of the earth body influences the amount of earth pressure. However, for a complete exact or numerical solution of the boundary value problem the time was too early, since plasticity theory for granular materials had not yet been developed and numerical calculations were extremely difficult.

Besides, Woltman's contributions to soil mechanics at the end of the 18th Century were far reaching and almost unknown in civil engineering circles. He not only introduced the angle of repose for the first time, which enabled him to create independently an earth pressure theory which was much more elegant than Coulomb's theory; he also developed some important ideas on soil mechanics and porous bodies.

Woltman (1794/99), harbor construction director (1757 – 1837) from Hamburg, first classified different kinds of soils. He separated soil into four types: sand, lime, clay, and compost-earth. He then pointed out that friction of all earth kinds is common and that this differentiates soil from fluids. Only by total saturation with water can the friction effect be lost. Referring to this state, Woltman called the mixtures quicksand, drifting sand, mud, bog, or morass. Furthermore, he accepted the incompressibility of soils, whereby he pointed out that this is most true for sand. Furthermore, Woltman regarded cohesion as a result of *mixed-in moisture*: "Dry and fine, the clay is dusty; if it is moist it stays together. Contrarily, clay has the special hygrometric property that its volume increases with additional moisture and decreases by drying, and that, after becoming totally wet, it becomes a rather solid mass during the subsequent **drying**. However, it must be borne in mind that, if the drying occurs quickly, then the contracting force can become larger than the cohesion. Consequently, clay tends to be broken up by many cracks. Sometimes **frost** can increase the cohesion of the earth masses, and can sometimes completely stop it."

After he had developed his earth pressure theory he checked his results by means of exceptions. He finally arrived at the observation that he could trust his theory: "I consequently believe that the formula cannot be doubted as far as the theory goes."

Woltman then communicated his "experience with earth pressure." He described in detail the testing procedure (set up by himself) and then imparted his test results for various granular substances. The care with which Woltman carried out his experiments is impressive. In particular, he listed all of the influences which might have falsified the results. He thereupon compared the test results with those of his theory. He subsequently showed that the earth pressure resultants, which, according to his theory, lay at one third of the height of the retaining wall, also approached this height in the test results. With respect to the size of the earth pressure, Woltman ascertained that the test values were much smaller than those calculated in his theory. He pointed out that his test results had been falsified by the friction on the walls of his box and therefore had to be corrected. He arrived at the following assessment: "Whatever the case may be, at this time I believe more in the theory than in the tests."

After strongly criticizing the test results of Delange, Woltman turned his attention to partially and totally water-saturated substances. He pointed out, in particular, that the moisture content augmented the cohesion but did not, however, attribute any greater importance to the cohesion. He attentively observed totally water-saturated loose soil, or mud, and noted that his theory was the most appropriate, since mud behaves like a liquid mass. In this connection, Woltman spoke of a mixture and, surprisingly, introduced the concept of the volume fraction. He was probably the first scientist to use this concept.

"Mud results from all loose kinds of earth if they are completely saturated with water or are in surplus with that mixed up. Its weight depends on the more or less admixture of the water.

We will assume that the empty spaces between the loose earth, which are filled with water, are to the bulk volume of that earth as $r : 1$, where r will be a proper ratio. The weight of the loose earth P; the water p, the volume of the earth v; of the added water u, then the weight of the mixture is $vP + up$; and its volume $v + u - rv$; therefore the specific gravity of the mud $= \dfrac{vP + up}{v + u - rv}$. If the mud does not contain more water than the spaces between the loose earth can hold, then $u = rv$, and the gravity of the mud $= \dfrac{vP + up}{v} = \dfrac{v(P + rp)}{v} = P + rp$ is largest. If, however, more water is added than the spaces between can contain $u > rv$, the mud becomes lighter if $P > p$; remains unchanged if $P = p$, becomes heavier if $P < p$. The latter can only occur for moor or peaty mold; I will soon add a table which can give more light hereupon."

Obviously, Woltman did not continue his interesting studies on porous bodies. There is no hint in any of his publications that he followed up on his original work on water-saturated porous solids with further studies.

After his remarks on partially and totally water-saturated loose soil, Woltman stated some design principles for dams and walls. He warned against bringing cohesion into the formulas, because this could be disconnected by becoming soft through frost and rain. In addition, Woltman gave some ideas for the construction of brick retaining walls, and addressed the influence of additional live loads behind the retaining walls and of the wall friction (soil-retaining wall) on the calculation. Woltman concluded his excellent treatise by pointing out that, for the achievement of a perfect theory of retaining walls, further investigations would be necessary.

In order to widen his knowledge in science and dike construction, Woltman had asked for leave in 1793 to continue his academic studies and to take an educational trip for a year and a half. After he had studied for one semester in Göttingen he set out on Easter 1784 and rode through Germany, France (Paris), England (London) and The Netherlands. He described the extended trip in sometimes lavish language in his four volume work (see Woltman 1794/99).

Until 1810, he developed his main capabilities in Ritzebüttel, in practical as well as scientific and literary respects. The work of Woltman's predecessor had been accompanied by many failures. Dike construction in Ritzebüttel had been carried out since 1733 with varying success. All constructions built up until 1740 were lost. The period from 1740 to 1756 saw the most solid constructions being erected, constructions which, apart from repairs, lasted for one hundred years and more. However, because these constructions were mostly built from stone, they were so costly, that for economic reasons they could not be continued. Therefore, fascine constructions took over. However, this construction method was found to be even more expensive. This was a result of the fact that the fortifications, which were made up of bushes and stakes, broke constantly. Thus, fascine construction was abandoned in favor of stone construction, although the procedure used had not yet been perfected.

The failures of the previous thirty years and the constant change in the construction systems used for the building of embankments came to an end when Woltman took over the direction of the dike concerns. He laid a sure foundation for protecting the works of Ritzebüttel, which, although they had been fully extended and built out, had been only minimally maintained for a long period of time. Woltman was able to achieve these excellent results owing to the introduction of, and strict adherence to, a few principles recognized as essential, which aimed at the greatest possible simplicity and strength of the constructions. He founded dike building on a few irrevocable principles. Woltman's great talent for abstraction was here made most evident; he had already proven this before (see his earth pressure theory).

One recognizes from the study of Woltman's career and his achievements that there were already useful ideas on soil mechanics and foundation work long before von Terzaghi entered the scene. This is not surprising, if one realizes that, for example, Venice and Saint Petersburg are both founded on extremely bad ground and yet they still exist. This could only have been performed by qualified engineers with a lot of practical and observational knowledge.

Moreover, there were other important contributions to soil mechanics and some related fields in the 19th Century, as e.g. O. Mohr's (1871/72) development of a failure condition for brittle and granular materials, Reynolds' (1885) investigation of the dilatation problem (the phenomenon of rising of dense sand layers under shearing stress) and, in particular, Stefan's (1871) creation of the mechanical mixture theory. It seems that Karl von Terzaghi missed these fundamental works completely, although he later became acquainted with these problems.

Karl von Terzaghi's contributions to the earth-pressure-theory in 1919 and later contained little mechanics but instead a lot of geometric considerations – and were very wordy. Also his investigations of the constitutive behavior of unsaturated sand, also new, remained unsatisfactory. His remark that sand behaved non-linearly elastically does not hold. The main deformations are plastic, which means that after unloading a sample deformations remain whereas in the case of elastic behavior the deformations vanish after unloading. Many researchers are even of the opinion that for sand the elastic range is so small that it can be neglected. It is amazing that von Terzaghi left the plastic range out of his investigations, although he mentioned enduring deformations, the more so since main aspects of plasticity theory had already been developed by Lévy, de Saint Venant in the second half of the 19th Century and, in particular, by von Mises in 1913 (see de Boer, 2000) with whom Karl von Terzaghi had been well acquainted since 1915 at Aspern (airfield).

Karl von Terzaghi's investigations on deformable saturated granular media were in some respects pioneering work as we have acknowledged. However, his descriptions of the stress and deformation states of saturated porous solids were not founded on a scientific basis. The partial differential equation which he developed locally describes the course of the excess of hydrostatic pressure in the pore-water in space and time. Moreover, it is only valid in consideration of very strong assumptions as has been shown by Heinrich (1938). Thus, it turns out from the above mentioned fields that Karl von Terzaghi remained in the very first stage of his efforts to develop a scientifically founded soil mechanics, namely in the stage of experience, observation and the recognition of problems or phenomena which he only proved through experiments. Without any doubt his great experiences in civil engineering, gained on building sites in Europe and the United States, his gift for observing and laying down his observations in written form and his keen sense for recognizing hidden physical effects brought him a deep insight into the mechanical behavior of unsaturated and saturated sand and clay. However, it becomes very clear from the aforementioned ideas of scientific theory that mere proof by experience is not sufficient for a long lasting theory. Rather one has to scrutinize a hypothesis or theory in view of its objective truth. That means that the hypothesis or theory must fulfill principles which are recognized as true such as the axioms in mechanics. Karl von Terzaghi's treatment of the saturated porous body violates the above statement in a flagrant manner. Although he had recognized that water-saturated clay was a mixture (see von Terzaghi, 1925b) he did not use the fundamental results of this theory developed by Stefan (1871) nor did he consider other elementary mechanical results, which had been recognized as true, in order to eliminate false theories or hypotheses like his theoretical study of consolidation.

At M.I.T. lecturing started in mid-September. Thus, there was a lot of time for travelling and relaxing. At the end of July, 1925, Karl von Terzaghi started from Constantinople for new frontiers. He rode with an acquaintance to the railway station. Some other acquaintances said fare-well to Karl. It was a beautiful morning. The beach in Floria looked like one of Turner's dreams of color. In the midst of the landscape a saffron yellow, lightened spot: that was Alida (an acquaintance, the author) surrounded by people waving (other acquaintances, the author). Via Budapest he travelled to Vienna where he spent an evening with Hofrat Forchheimer. From Vienna he rushed on to Graz, seeing Olga and Vera. Olga was "of chilly reserve as always."

In Karlsruhe, he visited the Technische Hochschule and Professor Rehbock and his labs. He observed several experiments.

On July 30 he left Vienna for Paris and on August 1 he arrived there at the Gare de l'Est, early in the morning. In the Rue de Strasbourg he found a small, modest hotel, though with a kind hostess. Karl enjoyed the typical Parisian atmosphere. Moreover, he was impressed by the Eifel Tower. "The construction was imposing for its daring and grace." In the evening he went to Moulin Rouge, on the Boulevard de Clichy. "Paris at night. On the broad boulevards bounded by rows of trees a flood of light and pulsating life ... – Moulin Rouge. A beautiful theater, almost as large as the Vienna Opera. One smokes, however one does not drink. No buffet. On the stage outstanding fireworks of light, colours, bright costumes or of nearly undressed female bodies."

After the show a carefully dressed man, obviously a stranger, asked Karl for the address of a certain nightspot. Karl suggested that they look for this place together and they ended up in an elegant apartment with marble stairs and paintings on the walls. "A folding-door opened and in front of us stood a dozen young girls. We took our seat on the sofa, music began to play and the girls began to dance in order to entertain us. A glass of champagne, we chose four out of the group and wandered with the four-leaf clover through some hallways to a hall of mirrors with a second broad sofa. There, the girls showed us with grace the thirty *positions de l'amour*. It was an amusing picture as the four supple, rosy bodies intertwined with each other twistings set into ecstasy through all thinkable contacts and contortions. Naturally the show ended in the tester-bed in the spacious bedrooms of the small good fairies, equipped with everything imaginable. The spectacle cost about forty dollars," a fair sum if one considers that at that time a dinner cost only about one dollar.

The visit to Versailles was unforgettable and the park made an overwhelming impression on Karl. "From the terrace the most wonderful silhouettes. To the right and the left of the broad middle field, English park, the statues lightly set into the green. In the right wing ring-wall with gates, in Rococo style. Obviously, the heart of the *Red Deer Park*. In the left wing, as a highlight, grotto of Apollo. In the middle of the very old park a rock wall appears which is reflected in the dark green water of a calm pool, overgrown with reed and water lilies. Halfway up the wall a grotto opens itself, against the dark background the bright figure of Apollo and of the nymphs stand out. I was always forced to stop and to admire the silhouettes. In front of the terrace of the castle, beds spread in bright violet symphonies of color of unbelievable beauty."

Slowly the time approached to leave Paris for New York. Karl von Terzaghi prepared carefully for the long trip and on Friday, August 7, 1925, in the afternoon he headed for Boulogne on the Atlantic coast with a luxurious train. He stayed in Boulogne for one night. In the evening, on the next day, at 7:00 p.m. he was on board of a steamer of the White Star Line bound for New York. During the next days he enjoyed life on board as usual with communications and entertainments.

In New York old memories returned: the view of the giant buildings, the loud noise of the trains and the atmosphere of concentrated business. During the next days he visited several acquaintances and he met George Paswell[31] and Lazarus White[32], noted engineers.

At the end of August, Karl travelled via Philadelphia to Washington D.C. in order to meet some authorities of the US Department of Agriculture, the Bureau of Public Roads and some other offices – and F.H. Newell, whom he knew from his first stay in the US. He learned from the authorities of the Bureau of Public Roads of the efforts to put road construction, which was booming as a result of rapid increase in car traffic, on a scientific basis via soil-physical investigations. The constructive criticism of these investigations which he expressed, already led after half a year to a fruitful cooperation which was to last for many years.

Finally, Karl von Terzaghi arrived in Boston, his final destination, at the beginning of September, 1925. He looked for an apartment and recognized that rooms in Cambridge were much prettier and much cheaper than those in Boston. He met Dr. Spofford and had a short conversation with him. To his surprise, Spofford told him that he expected von Terzaghi to lecture just

[31] George Paswell, noted structural engineer.
[32] Lazarus White, foundation engineer and renowned underpinning constructor.

one hour a week. During the next days and weeks Karl von Terzaghi became acquainted with his new colleagues and on Thursday, September 29, he gave his first lecture on foundation engineering to the students. In the following week he arranged his room, "brought some glassware and shelves into it, looks nice but no chance for work. No time – equipment must be ordered at once."

In October (Tuesday, 22) Karl lectured at the 20th Century Club in Boston. "Mostly old ladies and old gentlemen." His talk "Near East Recollections" was politely received. He reported of the beauties of Constantinople and the political developments in this exciting town and in Turkey, in particular, during the postwar time.

It was the consulting engineer John R. Freemann who paved the way for Karl von Terzaghi in the United States and, in particular, at M.I.T. between 1925 and 1929. He promoted him in every respect. Freeman bought some copies of *Erdbaumechanik* and gave them to renowned experts and wrote letters to several engineers and scientists in order to make von Terzaghi known in engineering and scientific circles. Some of the engineers and scientists had already read von Terzaghi's (1925 b) papers in the Engineering News Report and some had praised these articles to the skies. However, as already mentioned there were also some engineers who had expressed their objections concerning von Terzaghi's knowledge and some statements in the articles, like Thaddeus Merriman, from the Board of Water Supply, New York City, and son of Professor Merriman of Lehigh University. In a letter of November 12, 1925, he responded to a note of Freeman with hard criticism. "I have deferred passing judgement in this case because of the somewhat unfavorable impression which I received at the regular meeting of the Society on September 2, when there was presented a paper on *The Reinforced Concrete Arch in Sewer Construction.* At this meeting the gentleman in question presented a discussion severely criticizing these St. Louis constructions, but it was apparent to me that he was possessed of no practical knowledge whatever regarding such matters and that the suggestions for improvement of the design which he made were entirely impractical."

Moreover, T. Merriman stated, concerning von Terzaghi's first article in the Engineering News-Record: "It has been many a day since I have read an article so carefully as I have this one and yet have learned so little from it." And C. Barus wrote in a letter to J.R. Freeman on December 1, 1925: "I do not make anything out of the paper in question" (meant von Terzaghi, 1925b, the author).

Furthermore, in a letter to the Engineering News-Record, Professor Griffith of Iowa State College made the attempt "to advertise their work, to minimize mine and to excel in citations of many authors ... Wrote short and sarcastic reply. That man has not even read the text."

It was Hardy Cross, Professor of Structural Engineering at the University of Illinois, Urbana, who brought in a more positive tone in the criticism when he remarked in a letter to J. R. Freeman of December 27, 1925: "My first impression of Dr. Terzaghi's work in the News-Record is that he is going at a very difficult problem in a very scientific way ... I shall, however, go into his work more carefully as soon as I have a little time during the Christmas holidays"; and in a letter of January 27, 1925: "My general impression is that Terzaghi is doing valuable work." Freeman seems also to have acted as a go-between for President Dr. Stratton and Karl von Terzaghi, in particular, considering his salary, his laboratory and his reputation. In 1929 Karl von Terzaghi acknowledged Mr. Freeman's efforts in his farewell letter to the renowned engineer: "I beg you to remain assured that your unfailing constructive sympathy and your interest in my endeavors rank among the most inspiring impressions which I received during my presence in the United States, and I hope I will some day have an opportunity to welcome you in my Vienna laboratory."

Most of the time in November, Karl von Terzaghi dedicated to the preparation of his lectures on soil mechanics and foundation engineering as well as to proofreading a paper and investigating capillarity problems. In connection with the study of the capillarity phenomenon he spent much time reading a book by Freundlich: "Dip into capillarity-chemistry. Interesting and exciting."

It was at this time that he got in touch with the mathematician A. Ortenblad, who had just finished his studies at Harvard. Ortenblad had read von Terzaghi's (1925a) book *Erdbaumechanik*, and was not convinced by the analogy of heat propagation through an isotropic body with the consolidation problem. He started the development of his own theory in 1925, and presented it to the Board of Examiners as a thesis for a Doctorate in Science degree in May, 1926. In 1930, he published his thesis. In his thesis, Ortenblad (1930) formulated, for the first time, a consolidation theory for the three-dimensional stress and deformation states. However, he did not commence from the fundamentals of mechanics but rather merely extended von Terzaghi's differential equation.

A. Ortenblad became involved in an amusing affair between Karl von Terzaghi and the dean, as Ortenblad reported in an interview decades later. Von Terzaghi had started his lectures on soil mechanics, with subject matter well beyond an engineer's level, which proved to be too taxing for his students. Furthermore, he talked too fast. The dean requested Ortenblad to observe von Terzagh's course and report back to him. One day the dean visited the course and asked von Terzaghi to start the course again, owing to the fact that no one was able to understand the subject, and also asked him to teach more slowly, explaining in more details. Von Terzaghi became furious and said that there must have been a "spy" in class. Finally, however, von Terzaghi agreed to start the course again and he learned his lesson well. Later he talked slowly, made pauses after main parts in his lectures, explained the details, spoke in English only with a small accent, and he used rhetorical tricks to persuade his audience, like interrupting and confusing his opponent, oversubtilizing arguments, masking unpleasant facts with insignificant concepts as well as weakening proofs to the contrary by hairsplitting.

In December, 1925, Karl started the design of testing machines, even working on Christmas.

On New Year's Eve he received a letter from Professor Hans Reißner, who invited him to write an article for Winkelmann's handbook. Moreover, his efforts to set up a lab were making progress. "Thus the year 1925 ends with rather cheerful prospects for the future. During this year I passed practically through all the states intermediate between hopeless discouragement and the anticipation of seeing all the dreams of my earlier days realized. What I have pledged, on that summer morning in Berlin, 15 years ago, I will make true in Cambridge.

The whistles blow – announcing midnight. Blow very much more persistent, convincing and cheerful they do it, than last year, at the frontier between Asia and Europe! Instead of French Champaign – Whisky with Soda and instead of my Persian rugs cheap American stuff, but the Pioneer-work is done, and the harvest can start! My efforts have not been wasted."(sic).

At the beginning of January, 1926, he had lunch with Professor Max Born of Göttingen (Germany) and Professor Vallarts from the physics department at M.I.T. Max Born was a well-known physics professor and chairholder in Göttingen, later Nobel laureate, and also mentor of such famous students as Heisenberg, Jordan, Teller and Oppenheimer. At first Born made a "somewhat shy impression" on Karl, but grew rapidly in stature. "Very clean cut, characteristic profile, blue eyes, turned grey, carefully dressed, expresses himself accurately and carefully as people from Göttingen seem to do. Invited by Dr. Stratton, came probably September, leaves June 27 ..., who probably will rank with Bohr, Einstein, Planck and others. – Conversation with Born ... interesting. Effect of solid wall on viscosity. Effect of molecular collision, see fall of droplets, corrections of Stokes law. P.S. Epstein, Physical Review, recently. Diffusion, gas laws for dissolved bodies. Only mathematical conception, popular view. No reality: Vibrate, cannot more ... Kármán worked with Born 1912. Greater than Mises. Recently, theory on breaking of steelplates by electric sparks. Born continuously invited in company of physicists."

At the beginning of February, 1926, von Terzaghi urged Professor Charles Spofford to take action in his financial situation. Some days later Spofford spoke with President Dr. Stratton. However, the offer was not encouraging. The maximum that Dr. Stratton could offer him was a position of an Associate Professor at 3,500 dollars a year and probably some 5,000 dollars for the lab. "Hardly more ... I asked for time to think it over. However, on Saturday letter from Dr. Gates (president of Robert College, the author) that Robert College willing to wait another year." Thus, Karl von Terzaghi accepted the offer. Now, he was able to perform some small and simple experiments. He developed an improved ödometer. Moreover, he found time to write an article for the *Handbuch der physikalischen und technischen Mechanik* (Handbook of Physical and Technical Mechanics) (ed. by Auerbach and Hort) on the swelling of gels, and he intensified his consulting work. His income grew and he hired an assistant as well as a secretary. At the end of August, 1926, he noted in his diary: "The last four weeks were so thoroughly crowded with activities and events, that it seems they cover a one year's period of intense existence ... Last two weeks worked with increasing intensity on Chicopee and Westfield. Developed at this occasion technique to represent the facts. Got conviction, that the numerous mechanical analysises were wasted efforts. No laws where I expected such, computation of coefficients of permeability very unreliable. Future policy: develop method for approximately determining the coefficients of friction." (sic).

As has already been pointed out, in 1925, Karl von Terzaghi had published his *Principles of soil mechanics* as a series in the *Engineering News-Record* (see von Terzaghi, 1925 b). With this article he became known also in circles of consulting engineers and thus it was not surprising that consulting engineers tried to contact him in order to solve technical problems regarding dam constructions and foundations with his help. In the spring of 1926, he was asked to estimate the filtration loss which was believed to occur in a planned reservoir in Gramville, Massachusetts, and to design the foundation of a great pumping station in Lynn, Massachusetts, on quicksand and weak clay. The successful solution of both tasks drew the attention of experts more and more to the practical importance of the new methods in soil mechanics and the number of contracts grew steadily. The foundation of a 36-story bunkhouse in Houston, Texas, the investigation of the cause of the settlement of a large plant at the Pacific Ocean in the state of Washington, the construction of a barrage on quicksand in New England and the preparation for the construction of a hydraulically powered dam in Pennsylvania were particularly instructive.

At the beginning of the summer of 1926, Karl von Terzaghi was invited by the Bureau of Public Roads, Washington DC, as a permanent technical consultant to improve the soil physical test methods of the Bureau on the basis of the results of his own investigations and to reorganize the lab of the departments in Arlingston, Virginia. Moreover, one of the engineers who wanted to gain him as a partner was Daniel E. Moran. Karl met him in Spofford's office: "A lean, short man, with big head full of expression, sharp cut profile, hook nose, wide thin mouth, clear grey eyes. Looks as if he be not far from sixty. Contact was established before the first word was exchanged." Mr. Moran was a member of the Board of Engineers (with the excellent bridge constructor Ralph Modjesky as president) which had designed a bridge across the Mississippi. "They plan an open caisson foundation through top layer of very loose sand (river deposit)... Departed from Mr. Moran with the impression of having established permanent contact with a congenial personality." The Board of Engineers ordered him to perform the soil investigations for

Karl von Terzaghi at
Chicopee dam site, 1926

Karl von Terzaghi taking pictures

the foundation of the bridge across the Mississippi River. "In order to fulfil this task without neglecting my duties concerning the 'Bureau', I employed the young engineer from Vienna, A. Casagrande, as a co-worker, who was employed at that time as a draftsman by the Carnegie Steel Company in New Jersey." Arthur Casagrande later became a well-known engineer in soil mechanics in the US with much influence – and a disciple of Karl von Terzaghi.

Karl von Terzaghi had always been interested in the arts, in theater, music, and art exhibitions. Around the middle of the 1920s, he saw the great German actor, Emil Jannings, the first Oscar winner, in a theater performance in Goethe's *Faust*. However, he was not pleased by the performance: "A few splendid pictures (at outset, the plague, the old Faust in his studio), remainder unsatisfactory. Sentimental love story." Moreover, he was still enjoying the completely new field in arts and entertainment, namely silent movies. He saw with enthusiasm the "Potemkin film" staged by Eisenstein and the members of the Moscow Art Theater. Moreover, he was impressed by the movie *The Cabinet of Dr. Cagliari*. "Futuristic presentation of wild nightmares of the inmates of an insane asylum."

During this time he became acquainted with many Americans and people from other countries. As a single man he was invited by those people to lunch and dinner and he spent much time in their society, at the University Club, the Twentieth Century Club and the Cosmos Club. Professor Tyler and his wife took him to musical events and poetry readings. Moreover, Karl enjoyed very much the company of the German Consul Baron Kurt von Tippelskirch, a descendant of an old, noble family.

At the end of May, 1928, Karl von Terzaghi received a letter from his old friend Professor Schaffernak from Vienna with an announcement which could change Karl's professional position and his living situation completely. Schaffernak told him that Professor Halter, chairholder of Hydraulic Engineering II, was going to retire and Schaffernak asked him whether he would accept the professorship at the Technische Hochschule in Vienna. He was very pleased and happy reading Schaffernak's message, although he still hoped to receive an offer from Berlin for a chair in soil mechanics.

In June, 1928, Karl von Terzaghi was sent to Central America by the United Fruit Company, which needed soil-physical methods in order to be able to judge the grade of permeability of virgin forest soils. The goal was to learn whether the virgin forest lots could be converted to banana plantations by laying out irrigation ditches. At that time the United Fruit Company amounted to an American colony in Central America with its huge estates and many clerks.

The departure from the US to Central America occured in a great hectic state. Between June 6 and 9 he toured through the western part of Michigan in order to gain soil samples from the area on the Muskegon river. Back in Boston on Sunday, June 10, he prepared for his new adventure and two days later he entered the night express for New York. Next day, he embarked for his first destination, Costa Rica. On board he worked extensively on his report for his Michigan contractors. The steamer passed Cuba and made stop-overs in colorful Kingston (Jamaica) and Panama. Finally, on June 21, 1928, Karl von Terzaghi arrived in Port Limon, Costa Rica. During the next day he travelled to San José, where he was introduced to the president in the presidential office; moreover, he gave a talk on engineering geology (road construction and landslides) to about 20 students and he studied soil motion problems. From San José, he started a real adventurous trip on horseback to San Pedro and to the summit of Poas Volcano, a dangerous trip because it was the rainy season and the roads were "very poor." However, he had luck: "Wonderful view from rim of the crater: vast amphitheater, surrounded with pale colored, steep cliff, crowned by smooth or dissected dark green sandslopes. At bottom of the huge depression steam-jets rushing out, with noise like huge factory. The wind blew the smoke to and fro, occasionally exposing a small, quiet, elliptical lake of deep green color. On the right bank a mudflat, crossed by barrancos (canyons, the author) and cracks. Intense smell of H_2S (hydrogen sulfide, the author). In the rear and to left wooded hills, separated from crater by light colored sandflat. To left a terrific gap in crater rim, exposing onion like structure of lava and ash strata. From gap deep barrancos leading into troughlike depression 1 km wide, perfect desert (sic)." Back in the " 'hotel', intense rain started."

On July 1, 1928, Karl von Terzaghi returned to Panama in order to study large landslides along the Panama Canal – and to take part in new adventurous excursions. Karl was, in particular, impressed by the large Cucaracha slide: "Tremendous gap between two dark rockhills, the bottom lined with reddish clay, its rough, uneven surface strewn with boulders. Water flows over the chaos in small creeks and cascades, locally small pools."

During an excursion with some experts to other landslides and a new dam site up the Chagres River, Karl von Terzaghi experienced the real, exciting and sometimes scary world of the tropics with their snakes and crocodiles. On July 11, they started up to a mountain at 6:00 a.m.. "For two hours on well kept trail through jungle ... Native (Negro) banana plantations. No permission to open new plant, because expropriated for US ...

Waded P. River three times. Climbed up a steep hill, passed the foot of a huge, grey limestone cliff in the dark shadow of primeval forest. In a creek nearby a huge green snake, lifted her head two inches out of water, showed white underside, followed us with eyes as we passed by and descended into valley at tunnel entrance. Like landscape of earliest days of earth. Gigantic trees closed up above pool, with black cave in background ... Tropical thunderstorm, lightning flash and downpour ...

With the canoe at 2:30 at Chagres River ... About half a kilometer downstream ..., our canoes approached overhanging rock, and before I clearly understood what happened, the canoe was soaked under it, the boat capsized, and my Kodak slipped off of my arm and I started to swim. Followed outer rim of rock, until I reached a point, where climbing was possible."

On July 16, Karl von Terzaghi left Panama and sailed from Colon to Puerto Castilla (Honduras) and rode further by train to Trujillo. During the next days he was received by authorities of the United Fruit Company at different places in Honduras and he performed tests in the soil labs of the Company, e.g., in Tela, close to the coast between La Ceiba and Puerto Cuertés.

Karl von Terzaghi's next destinations were Puerto Barios and Quirigua, one of the most important Maya towns, located in the east of Guatemala and known through large altars and tall steles. From Quirigua, Karl started for sight-seeing tours to Chiquimula and of course, to Guatemala City. From there he rode to the border of San Salvador and to Antigua, the marvellous former capital at the foot of the powerful volcanos Agna, Fuego and Antenango.

On Friday, August 31, he decided to ride by train to Quirigua, because he felt ill. In the hospital he was inspected by an acquainted physician who stated: "Heavy inflammation of inguinal glands, and amoebic infection of upper intestinal tract, ..." He stayed for two weeks in the hospital and rode then to the Gulf of Honduras with a motor boat and embarked for the United States.

After his return from Central America on Wednesday, September 26, 1928, Karl von Terzaghi found two pleasant surprises in his mail. First, he was asked whether he would accept a position as director of the *Versuchsanstalt für Wasserbau und Schiffbau* in Berlin. Second, he was requested by the Japanese and American Committee of World Engineer Congress to speak in Tokyo in October, 1929.

The four-month-trip to Central America had a great influence on his daily work. "With doubled joy and interest at my work." This was quite necessary, because he had to lecture and do consulting work. He was requested to clarify the causes of unexpectedly occurred settlement at a large plant on the coast of the Pacific Ocean, close to Canada. Only two months later he was asked by the state of New Hampshire to review the calculation documents for the lay out of a retaining wall, 55m high, at the Connecticut River. "This task could only be solved with the help of earth pressure tests on a large-scale and the great amount of building expenses justified the acquisition of a large earth pressure apparatus, equipped with all auxiliaries of the modern measuring technique." In a relatively short time the large devices had been built and installed. The tests lasted five months and the results gave a clear picture of the phenomena caused by the movement of the retaining wall as well as of the combined actions of earth pressure and water pressure at a weakly coherent back fill. "The data gained answers all open questions and supplied a reliable basis for the design of the planned wall."

On New Year's Eve, Karl von Terzaghi reflected on the past year. "Last day one of the richest years of my life shaped by numerous rapid decisions ... The last weeks the great retaining wall adventure. Looking back on this year, I find the boldest dreams of my early manhood by far surpassed. External honors also came, by and by, during this year: Medal of Boston Society, Fellow of American Academy of Science, Advance Sciences, full Professor, *korrespondierendes Mitglied des Österreichischen Ingenieur- und Architekten-Vereines*, Congress Tokyo, Calls for Berlin, Vienna. The only dreary spots between, the days on board ... the ocean in its immensity compensated for the dullness of the passengers.

Whistles blowing everywhere in Boston and the new year starts, a blank space, to be filled with the creatures born out of our imagination fertilizing circumstances. What shapes and pictures are going to grow in it? I greet it with a glass of rum, rescued from the shores of Guatemala.

My satisfaction: I give five times more than I take and earn five times more than I need."

After his reflections, he described some recollections in his diary concerning the last five months of 1928. In November, 1928, Karl's assistant, Gilboy[33] (he served later on the Civil Engineering faculty teaching soil mechanics until 1937), had mentioned an interesting conversation with the president of M.I.T. "Dr. Stratton: 'You work for three years, and no result!' Gilboy: 'Dr. T. most excellent results in his consulting work!' 'Yes, this consulting work. I don't like that at all. Distracts from fundamental research.' " Indeed, Dr. von Terzaghi neglected fundamental research to a great extent in favor of his consulting work and his scientific findings were more than meager during his time at M.I.T. and in no way comparable to his work in Constantinople.

And he mentioned another recollection of a meeting which was to have a lasting impact on his personal life. "Back to 1928. December 18. Phonecall from Miss Doggett. Geological Society, for professional advice. Met her first at geological excursion on September 12, quiet, serious, yet with occasional flashes of original humor. Called on my office at 5 p.m. Thesis coral reefs near Chicago. Excessive steepness of coral sand slopes. May be due to unequal compacting. Dinner at University Club. A pair of eyes which penetrate, and shine in a face with extraordinary variable expression." This young lady who had obviously impressed him very much was Ruth Doggett. She was a descendant of an old family of European immigrants with some roots going back to the first immigrants in Boston. Ruth remarked in a letter to the author on July 20, 1990: "My father's first ancestor of the same name (Doggett) came to this country in 1630, as a bonded servant, i.e. he was required to serve at little or no wages to work off the cost of his passage. Not quite 'an old English family'. At some point, a male Doggett ancestor married a decendant of Dr. Fuller, who arrived on the Mayflower along with the first settlers in New England, in 1630 (the Puritans). We would thus have the qualifications

[33] von Terzaghi's doctoral student

for belonging to the Mayflower Society, composed of a bunch of conservative, stuffy people with whom I would not wish to associate." Her mother lived in California. Ruth's maternal grandfather came from Rhineland, Germany, emigrated to the US in 1848, where he ran a pharmacy. Her mother's mother was from Kentucky.

On January 24, 1929, Karl had dinner with his old friends Mr. and Mrs. Tyler at the University Club. Tyler and an acquaintance "made effort to start action for keeping me in the US. Yet Spofford, with his department pride, opposes any action which does not pass through his hands."

However, there were also great efforts in Vienna to get Karl von Terzaghi to the Technische Hochschule. On February 5, 1929, he wrote in his diary: "Great joy: Letter from Austrian 'Aktions Kommitee', headed by Dr. Exner[34]. Offer to me the means for establishing a soils laboratory in Vienna. What a gorgeous satisfaction, after the guerilla warfare with Stern and Co. That would be a beautiful farewell from Cambridge. Quit, when everything seems to be most promising."

After Professor Max Born from Göttingen had visited M.I.T., it was his famous student, Professor Werner Heisenberg, at that time only 27 years old who lectured at M.I.T. in March, 1929. On March 15, "At 4 p.m. First M.I.T. lecture of Dr. Heisenberg on Quantum Mechanics. A boy, 24 or 25 years old, slightly awkward when facing his audience. Full, healthy face, good color, grey eyes, nothing extraordinary in his expression, except, probably,well developed forehead. A boy, who enjoys the good things of life. Lecture in big lecture hall, some 350 attendants. Explained his topic in a fairly pleasant way, as more or less incoherent comments to his slides. His further lectures a success, attendance kept up to 80." Werner Heisenberg, one of the great men of modern physics, received the Nobel Prize in 1932.

During springtime and summer (1929), the relationship to Ruth Doggett got closer and closer. Very often they had lunch and dinner together, visited concerts and theater performances and spent the weekends with hiking and sight-seeing. Ruth's mother was concerned about the relationship. She spoke to Karl: "On that Sunday when you had dinner in our home, I saw the threatening sign on the wall. Sunday, yesterday, when Ruth came home, my worst anticipations were confirmed. You became with her an obsession. She confided to me her passion, and I had a sleepless night. No means. When she is through with her studies, the money is gone. You know how rigorous it is in US about morals ... She is young ... Will you promise to me that you remain – just friends? You are a man of the world. Can you help me to protect the girl?"

However, Karl was not really bothered by the concerns of Ruth's mother. He was much more disturbed by the lack of information from Vienna. And at

[34] Dr. Exner founded, equipped and ran the Technologisches Gewerbe-Museum in Vienna for 25 years and was a well-known engineer with much influence in engineering circles in Austria.

Ruth Doggett (von Terzaghi)

the end of May he noted in his diary: "After several days of indecision made up my mind to stay in Cambridge. Called on Spofford, requested salary of 6,000 – and title of Research Professor. Spofford unusually agreeable. Convinced that I have greater future in America. Will take matters up with Stratton at once. There will be no difficulty to give me leave of absence for several months, even for half a year, to go to Europe or to take up ... work like that in Soviet Russia. One hour later I received a cablegramm from Vienna: 'Interministerielle Verhandlungen noch nicht abgeschlossen'. This is, as it seems to me, my lot to stay in the US."

Some days later, Karl von Terzaghi had a one-hour conversation with Mr. Spofford who told him that President Stratton wanted him to drop consulting work. Also J.R. Freeman complained that he was "too much leaning towards commercial side." Karl replied to these accusations in his diary: "Spofford agreed that Freeman does not live up to his requirements and that there are many members of staff who could not possibly earn outside of M.I.T. more than they do here. Spent the second part of afternoon frankly explaining my reactions and presenting my ultimatum."

Then, finally on Tuesday, June 5, Karl von Terzaghi received the long awaited telegram from Vienna with the news that practically all conditions had been accepted. "Although the telegram was to be expected, it left me in a state of surprise and I postponed reaction. Intended to use it as means to secure second assistant, and then to wire back 'too late'."[35]

He spent the evening with Ruth. "My passion for the little girl still growing. Passed through hours of most intense emotions." Gradually, he was thinking about a closer relationship.

On the next day, he spoke with Ruth's mother about Ruth's future. "Told her of faint possibility of various intentions on my part. Her financial requirements? Friends in Chicago trusted with tiny capital. If she does not live too long it may last. Would feel mortified, if she depended on outside support. Husband – Irish – grand style – generous – lived beyond his means – died prematurely ... work manager of advertising ride of big firm. Daughter in California, artistic, sunny disposition. Married brilliant lawyer against father's will. He gave up profession, small writer, family supported by daughter Marguerite. Excellent family, but unbearable conditions. Later, matter of fact tone displeased me utterly, left disagreeable taste."

Karl von Terzaghi was also dissatisfied with Spofford's attitude in the matter of the assistant which was a growing confrontation. He spent a great part of the night pondering his decision. He went desperately back and forth. "Ruth figure representing additional disturbing factor. On one hand great possibilities for activity in US – growing reputation. Time consumed in Europe in actual teaching, depressing atmosphere. Other hand: deadly uniformity of life in the US. No change for escape from demon economies ... Humiliating attitude of Dr. Stratton. Wealth of new contacts in Europe. Then I reached my final decision."

On June 7, 1929, at 2 p.m., Dr. techn. Karl von Terzaghi handed in his resignation to the head of the department, Professor Spofford. He received the resignation very ungracefully. "You will feel sorry – all people whom I spoke said a great future ahead of you – we thought everything was cleared up. Present your letter of resignation."

He spent the evening with Tyler's Jewish friends and with Ruth on a lounch in the Boston Harbor. It was a farewell party for the Tylers who were moving to Washington where Professor Tyler had been appointed as Dean of Engineering.

"When I came home, tears were flowing from my eyes. I felt what I am going to lose." On the next day, his mind was still in a state of revolt. Finally, however, the atmosphere cleared up. "If passion still redhot and no hope for substitute, then go ahead and marry."

---

[35] The personal conditions were not bad. His salary was more than 18,000 Schillings per year plus lecture-fees plus examination-fees plus a personal bonus of 15,000 Schillings. In addition, the minister promised to credit fifteen years for his pension. This was a maximum salary corresponding to the salary of a general. One can imagine how high his personal extra pay was if one compares it with the extra pay of his colleague Professor Lechner, namely around 6,000 Schillinge per year.

During the following time, Karl von Terzaghi said good-bye to some friends and acquaintances in Boston, New York and Washington DC. The love between him and Ruth grew steadily and he introduced her to his friends who accepted her in a very friendly way. Mrs. Tippelskirch: "We were quite enthusiastic about Miss Ruth. Very lively, very clever, and what is more, she has personality. Yet: Do you have the feeling, that she is the only one?"

At the end of September and the beginning of October, 1929, the days were filled with strenuous work in Cambridge. "Finished pile report, prepared report on Hartford water supply and last retaining wall tests. Saw Ruth every evening for a few hours. She helped me packing. October 1 until October 2 at 4 a.m., finishing my bags."

On his birthday, October 2, Baron von Tippelskirch and other friends gathered at Karl's office while he wrote the last pages of his report. "Last, cordial farewell." At 4 p.m. at home, Ruth prepared tea and helped him to get over the effects of the last two strenuous days. "Between passionate embraces, a few serious words about the possibilities of marriage ..."

"At 11 p.m., I had to leave for the railway station." It was an emotional good-bye from Ruth.

Thereafter Karl reflected on his birthday "... feeling of having experienced the most significant birthday of my life, in a certain sense the climax of my career: External success beyond my boldest dreams, earnings in 9 months 23,000 dollars. International reputation firmly established, and the devoted love of a courageous girl with brains and imagination."

On October, 4, he had a brief lunch with old friends, among them Paaswell, White, Gilboy and Casagrande. "Told White frankly my opinion about his underpinning book. He became chairman of Soils Committee and I suggested cooperation with European Committees. Gilboy, during last week failed to prepare settlement tracings and failed to check figures of pile report. Abominably lazy, lets himself go carelessly.

Evening dinner Engineers Club with Melzer, Paaswell, Schmitt and the boys. Happy, cordial atmosphere, with Scotch and ..." Schmitt asked him: "Four years in US. What are your reactions as to Americans?" Karl told "frankly about extraordinary open mindedness of profession, of active, interesting men I met. A. Schmitt listened silently."

Karl rode with Meltzer's car to Brooklyn. "Gigantic dock, crowded with people. Liner, like a huge hotel. Casagrande, faithfully, attended to baggage. In my room, large like a hotel room, special delivery letter from Tippelskirch, letter of introduction to ambassador in Moscow and assurance of his friendship. Two beautiful letters from Ruth. After the crowd had left, for an hour with Casagrande in the Wintergarden at near end of boat. Surrounded with luxury. Half dark, big, beautiful groups of flowers and palms, small tables between them. Appropriate setting for retrospective contemplation of the past years. C. reminded me that 3 years ago I doubted whether I will stay. He himself still one year ago, was not sure of future. Now he is. Whatever he became and he is, he owes to me. During last year, self confidence increased. I called his attention to his expertship in soil testing. 'No chance for us, unless we are dragged on and up by mightier spirits. I have passed through this stage. Now it is easy. I will never forget what you did to me and I feel the deepest gratitude for it'. This conversation was the most beautiful closure of a four years strive, started as a single handed one –."

Then, Karl von Terzaghi continued the description of the interior of the steamer *Bremen* and life on board. "First day on board extremely dull. Mostly ... voyagers on a larger scale, much noise and little grace on board. Interior decoration in part very beautiful. Vast social hall, dark smooth brown lining throughout, large, simple shining brass ornaments between roman, arched windows. Smooth, brown, slender columns with brass stripes. Entrance to ballroom both sides fantastic marine motives in mosaic. Exit from dining room: transparent columns, simple copper and brass Roland statue. Yet impression of a household up to the limit of income. Lift handly functions. Incorrect information. No booklets describing the boat. Called at 5 p.m. on captain. Big, fairly stout, round-headed, red-haired man with rosy cheeks, fairly jovial. General talk. Instead of 'Kreisel' rolling tanks with oil in it. Very economical machinery (small consumption of fuel oil, record)... Asked permission for Casagrande to come ... Main dining hall. Space for 670, two launches, china room, completely paneled with plates, hunting room, paneled ceiling, walls with hunting scenes (Middle Ages) painted on linen."

Karl also took part in social life on board. "Yesterday cocktail party in the captain cabin. Some ten gentlemen, various description, mostly big industrialists. My neighbor, 'general manager of boat' (nearly 1,000 employees) since many years mostly the far East. Cold atmosphere, no conversation, everybody sitting stiffly in his chair. My neighbor, roundheaded German, head of Asbestos distributing organization."

Karl used his free time extensively for working out reports and for finishing his correspondence – and he thought often of Ruth. "If I ever experienced the beautiful sensation of sharing with a human being all my thoughts and feelings and finding perfect response, it was with you, during last three months. With you, my birdie, the strain was eliminated and all my latest capacities for enjoying life celebrated resurrection. Whatever the future may bring, never will I forget the love which surrounded me during three months like a precious atmosphere."

Gradually, the *Bremen* approached Europe. The final two days on board were filled with music, dance, and orgies of intoxication.

On October 10, the *Bremen* arrived in Cherbourg and Southhampton. The next day the steamer approached Bremerhaven. During the night there had been a heavy storm and the water was dirty grey-green and very rough. In Bremerhaven they waited four hours for Baurat Agatz, who had promised to receive them. After visiting the harbor and a lab they rode by car at their own expense to the railway station. At 9 p.m., they arrived at Bremen's main railway station. Through the main street they came to the public place of the cathedral passing dark waters – and visited the *Bremer Rathauskeller*[36]. They left around 10 p.m. At the railway station Karl said good-bye to Casagrande, who wanted to go directly to Vienna, whereas Karl rushed for Berlin where he arrived on Saturday, October 12. He was lodged in the hotel Eden where he felt at home. During the next days he visited several people and acquaintances, for example the successor of Professor Krey and Professor von Mises. Krey's successor seemed to be a failure. With Richard von Mises, "Professor at the University, prominent mathematician, and connoisseur of the arts," he had dinner. He told Karl that von Kármán had gotten a position at Caltec in Pasadena, California, where he was working for a considerable time with great success in the field of mechanics.

On Sunday morning, Karl von Terzaghi visited an art exhibition with prints. He was impressed by the facsimile of Gauguin's *Noa Noa*. In the afternoon he went to the National Gallery. "Splendid Böcklins – Friedrich – Bracht, Marie. The old friends."

Shortly before his departure for Europe Karl von Terzaghi had been asked by the Soviet Government to prepare an expert opinion on locks built within the Volga-Don-Canal project. He agreed immediately. He left Berlin on October 14, at 11 p.m., with one of the airplanes of the Luft Hansa accompanied by the finance attaché of the Soviet Embassy. "The cabin of the plane resembled the interior of an old-fashioned, wornout autobus, the seats crowded together and the sleeping facilities were limited to a belt arrangement, to be fastened around the waist." The nocturnal flight was something magnificent. Berlin gradually turned into a dark carpet, covered with millions of lights. At about 3 p.m. the plane had to make a stop-over in Königsberg. A primitive restaurant served on the aerodrome as a shelter, reminding one of the "Schutzhütten" in the Alps. "In Königsberg we had to wait until sunrise, because beyond Königsberg the route is no longer staked out with light signals." Half an hour after they had started again the plane came into a heavy storm, the most violent Karl had ever experienced in the air. Within a short time, the stomachs of most of the passengers "were entirely empty. An Englishman suffered a heart attack and looked as if he was going to die ... I myself remained unaffected merely on account of the enthusiasm I derived from observing the skilful manoeuvring of the pilot ... Nevertheless, I felt greatly relieved when Riga appeared on the horizon." Karl stayed with his Soviet companion in the Hotel Metropol, in the same one he had stayed in in 1911, when he was working in Riga, "once the scene of high life, and now as quiet as a cemetery after a funeral." Most of the time, he waited at the aerodrome expecting better weather. On October 17, they had to make up their mind to travel to Moscow by railroad. At the Russian border all their documents were carefully scrutinized and the newspaper, which Karl had taken with him from Riga, was partly confiscated. The Russian trains were in bad

[36] Famous old wine cellar with a collection of excellent wines

shape. "Nowhere, except in the remoter parts of the Balkans, and of Anatolia, have I seen trains in such a state of neglect." The various train-stations were almost "crowded with peasants and beggars and women with babies, the faces pale and haggard, no hats, no white collars, the dresses mostly torn and dirty." The crowd "seemed to have escaped from Gorki's night asylum."

From Moscow he rode to the construction site of the planned channel. The Russians planned to build the channel within the next *five-year plan.* Nine locks were needed on the Volga side and three on the Don descent. The generous experiments, which the Soviet-Government had commissioned on the plateau between Don and Volga, gave the hint for a study by Karl von Terzaghi about the judgement of the filtration lost in channels.

The Soviets did not hesitate, lacking their own experienced engineers, to hire the best experts in the world at great cost – and Karl von Terzaghi came back to Russia the next year.

Karl von Terzaghi was appointed o. ö. Professor (full professor) on November 1, 1929.

In Vienna, a lot of work awaited Karl von Terzaghi. He had to set up the chair of Hydraulics II, to prepare his lectures and to install the devices for his experiments. Arthur Casagrande was a great help to him, as he took over almost all the cumbersome work of equipping the lab. However, despite his professional work he found the leisure time to receive and visit old friends, like Mizzi Obermayer and Hilde Leslich. Hilde Leslich, with the maiden name of Uray, was an artist. Karl mentioned in a letter to his mother that Hilde might receive an offer for the decoration of churches in the Ruhr area in Germany. Besides, Hilde Uray (her artists name) later shaped a bust of Karl von Terzaghi.

Despite all of his activities, Karl very often recalled the good time he had spent with Ruth the last months before his departure to Europe. His desire for Ruth grew steadily and, finally, he decided to marry her. "My success in my public career should bring up also a success in my purely human relations and I believe to know my coming wife just enough in order to be sure that my expectations will be fulfilled." However, it was not so easy to perform his plan. Ruth Doggett was still occupied and busy in Boston. Therefore Karl decided to marry her by proxy. "After endlessy running about from office to office in Vienna and telegraphic communication with Cambridge I have collected all the documents needed for the marriage." He married Ruth on Monday, June 2, 1930. Mrs. Lotte Reskiwer acted on behalf of his fiancée. Around this time Ruth was already on her way to Europe after she had received her doctor's degree. She embarked in New York on May 30 and arrived in Bremerhaven on June 5, and on the next day in Vienna.

In the summer of 1930 Karl was asked by the Swedish engineering company, Vattenbyggnads-Byran of Stockholm, to return to Russia. On July 21, 1930, he travelled with Ruth via Berlin to Stockholm. During the next days they did a lot of sight-seeing. On July 30, in the evening, they headed to

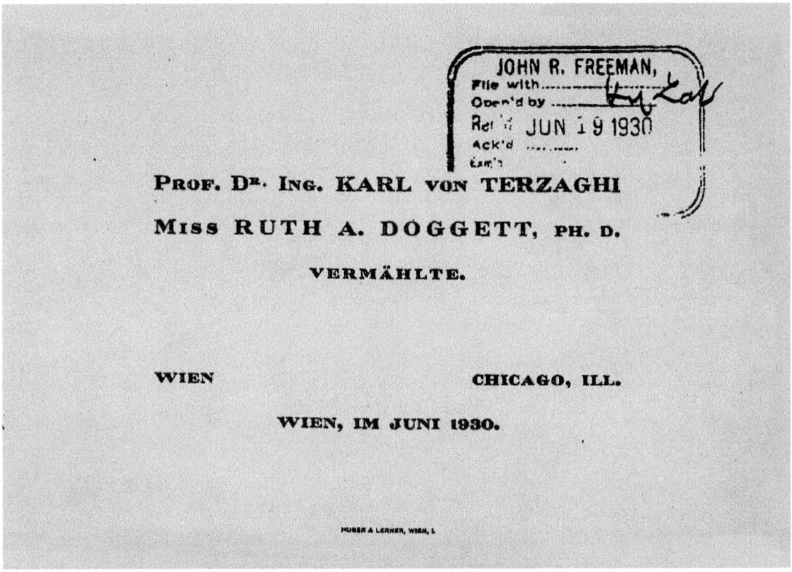

JOHN R. FREEMAN,
File with
Open'd by
Rec'd JUN 19 1930
Ack'd

PROF. DR. ING. KARL VON TERZAGHI

MISS RUTH A. DOGGETT, PH. D.

VERMÄHLTE.

WIEN CHICAGO, ILL.

WIEN, IM JUNI 1930.

Marriage announcement

Helsingfors with a steamer. On the next day at 3 p.m. they were received in Leningrad (St. Petersburg). Karl was very pleased to see all the points of interest in Leningrad which he was familiar with from his first stay in this exciting town in 1911. In the middle of August, 1930, they arrived at a construction site on the Swir River, between the Ladoga and the Onega Seas, where he spent several weeks. Karl had to examine the foundation of a large reservoir on a mighty layer of stiff clay with thin, plastic intermediate layers and embedded sand with artesian water. This was a good opportunity to compare the calculated settlements with observations. He finished his work on August 15.

Next he was asked by the Soviet Government to examine an earth dam in Bobriku, south of Moscow, which was under construction, along with some dam projects in the immense canyons of the Sulak River in the eastern part of the Caucasus Mountains and an interesting hydraulic power project in the Armenian highland.

Via Moscow, von Terzaghis returned to Leningrad where Karl formulated his report and organized meetings to discuss the problems of the various foundation works. Moreover, he used the opportunity to go to Stockholm, where the 3. International Congress for Applied Mechanics was to take place in the Parliament House. The Congress was opened on August 25. "In the evening, banquet in the marvellous meeting hall of the municipal house. Around 20 or 30 meters high, flowers, champagne, walls covered with mosaics. Ceiling high above the festive tables. Flourish of trumpets announcing the speeches." Karl von Terzaghi met many old friends and famous scientists, e.g. Wester-

A pile-load test, Rotterdam, 1930

gard, von Kármán and Fellenius. On Friday, August 29 – the last day of the conference – he gave his talk in the Parliament House in front of an audience of nearly 50 scientists. A contribution on his part is not contained in the proceedings; maybe he did not send in a paper.

After finishing some reports in Stockholm he returned, together with Ruth, to Leningrad, where they stayed for several days, before heading for Tiflis. They arrived there on September 21 and Karl was very busy with consulting work over the next days.

On October 8, in the evening, they stopped over in Constantinople after a long ride and, finally, returned to Vienna.

In the summer of 1931, Karl was called to North Africa by a firm from Milan in order to study methods for making the foundation of a 50-meter-high dam safer. On August 4, in the evening, he departed from Vienna. He stayed for three hours in Venice and then rode to Milan, where he met with Mr. Rodio, who remained a friend of Karl for a long time. Via Genoa and Toulouse he made his way to Marseille where he embarked for North Africa. On August 11 he arrived at Oran and on August 13, in the late afternoon, at his destination, Bou Hanifia. He stayed there until September 3, 1931, and returned then to Vienna.

Karl von Terzaghi was thinking about returning to the US in 1931. He wrote to the renowned civil engineer Lazarus White who arranged contacts

Technische Hochschule in Vienna

with Dean Baker from Columbia, New York, concerning a position for von Terzaghi. Upon an offer of Dean Baker, Karl applied a full time research professorship. The reply of Columbia is not known. However, his application obviously failed, because von Terzaghi stayed in Vienna until 1938.

The year 1932 was not a very productive one. Karl von Terzaghi spent most of the semester vacation in North Africa – Bou Hanifia, Tanger, Fez – at the Riviera (Monaco) and in the Dolomites (Cortina). In Bou Hanifia, he continued his consulting work and in the Dolomites he relaxed with hiking and climbing. The tours were dangerous in part and could only be performed with the support of a mountain guide.

At the beginning of the winter semester 1932/33 von Terzaghi reduced his consulting work. "Strenuous work on paper for the Second World Power Conference. New insights, formulating of the term 'Rutschneigung'." Moreover, he managed to attend a Round Table Conference of the Degebo[37] in Freiberg, Saxony, Germany. Karl von Terzaghi was not satisfied with the conference, in particular with the poor organization of the society. "Already in the evening I had in mind to organize Degebo and to replace Früh (the chairman, the author) with Loos (his former assistant, the author). ... Loos is enthusiastic about the idea."

[37] Deutsche Forschungsgesellschaft für Bodenmechanik (German Research Society for Soil Mechanics)

Karl and Ruth von Terzaghi with institute staff, 1932.

Gradually he returned to scientific questions, in particular, to the uplift problem, which had been an issue since the turn of the century. And in this connection he mentioned the name Fillunger for the first time, chairholder of Technical Mechanics at the Technische Hochschule of Vienna. On December 10, Karl von Terzaghi noted in his diary: "Finally, investigations on uplift on barrages – unpleasant discussions with Fillunger, who wants to be praised and supported his awkward and unrealistic theory and stops at nothing in order to disguise obscurities and false conclusions. Work on the completion of my work without joy."

Karl von Terzaghi's disagreement with Paul Fillunger's results concerning the uplift phenomenon would become the root of one of the heaviest and bitterest disputes and biggest scandals in science history – with a very sad ending.

2. Prelude to the Scandal

2.1 Paul Fillunger – a Scientist and Critic

Paul Prosper Fillunger's roots are also embedded in the Austrian Empire and can be traced back to the 18th Century, in which his great grandfather, Johann Georg, was born on April 5, 1781. His grandfather, also named Johann, lived from 1807 till 1879. Paul Fillunger's father, Johann (1848-1917), broke with the tradition of giving one son the name Johann. Grandfather Johann was a well-known civil engineer and was involved in the development of the so-called *chain-bridge*. From 1842 on he was involved in civil service and supervised the construction of a railway in Austria. In 1856, he left civil service and performed various private constructions. There were several members of the family who had worked in the field of engineering. Thus, the course of Paul Fillunger's life, from his birth in Vienna on June 25, 1883, was already marked.

Fillunger family in the first half of the 19th Century
with his grandfather on the right in the front row

Fillunger successfully completed the four classes of the junior high school in Vienna from 1894/95 until 1897/98, and the three classes of the upper high school in the k.k. (kaiserlich, königlich) state-upper-high school in Linz from 1888/89 until 1900/01.

During Paul's childhood, Johann Fillunger married for the second time. Helene Fillunger was Paul's unbeloved stepmother and it was rumored that she forced Paul out of the house to live as a boarder in Linz. Maybe his sternness in later years had its roots in this occurrence, a sternness which should influence his life decisively later.

Paul passed his final examination with honors. In 1901, he enrolled at the faculty of mechanical engineering of the k.k. Technische Hochschule of Vienna. After studying for four semesters, he did some practical studies in 1903. In the same year, he passed the 1st examination (Staatsprüfung) with the degree *capable* and, after doing further practical studies, the 2nd examination in 1904 (Staatsprüfung) with the degree *very capable.*

In 1906, Fillunger entered the private Austro-Hungarian State Railway Company, where he stayed until 1910. During this period, he had some free-time for his hobbies. For example, in 1907 he passed the official examination for becoming an engine-driver and a tank-attendant. Moreover, he taught himself differential and integral calculus with enthusiasm, using E. Czuber's book. In addition, he widened his knowledge in the field of strength of materials, guided by books by C. Bach, *Elastizität und Festigkeitslehre*, by L. v. Tetmayers' textbook with the same title, and by G. Kaiser's textbook, *Construction der gezogenen Geschützrohre.* In 1910, Fillunger took over a teaching position at the Technologisches Gewerbemuseum in Vienna and, in 1913, he was awarded with the title k.k. Professor; his teaching areas were mathematics, mechanics, and mechanical engineering. His career was interrupted in 1915 when he was called up as a *Landsturmingenieur* (engineer of veteran reserve). In the last years of World War I, he did work which proved to be useful at the Fliegerarsenal, in Aspern.

After World War I, Fillunger was appointed chairman of the laboratory of the Technologisches Gewerbemuseum. Fillunger worked successfully there and, on September 9, 1920, he was honoured with the title *Baurat.* In 1923, he was offered the chair of Technical Mechanics at the Technische Hochschule of Vienna, which he subsequently accepted and received the title o. ö. professor (full professor). However, he was only able to take over the chair, and not the laboratory, which was incorporated in the chair of mechanical technology (Prof. Dr. Ludwik).

Fillunger started his scientific career in 1908, when he received his doctorate degree at the Technische Hochschule of Vienna. The title of his dissertation work was: *Ein Versuch, die Spannungsverteilung in keilförmigen Körpern auf theoretischem Wege zu finden* (An attempt to find the stress distribution in wedge-shaped bodies in a theoretical way). Fillunger's achievements were honoured with *excellence.*

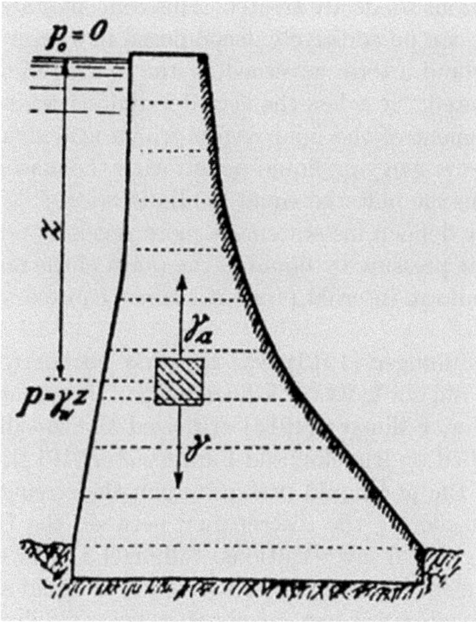

Fillunger's consideration
on the uplift in a barrage.

Despite his teaching duties at the Technologischen Gewerbemuseum he found time to do scientific work. Already in 1910 and 1912 he had reported on stress states in wedge-shaped bodies and the application in barrage construction in the renowned journal Zeitschrift für Mathematik und Physik (see Fillunger, 1912). These treatises were extended versions of his dissertation. The occupation with barrages inspired him, obviously, to go deeper into the problems of these important buildings, as some of them seem to have occupied him for a long time. The first phenomenon he investigated was the uplift problem. Already in 1913, he published an extended paper on this problem. It was Fillunger (1913) who pioneered the porous media theory of liquid-saturated porous solids with his contribution. The study of the analysis of the force acting in heavy-weight masonry dams had led him to this problem. He developed a simple formula for the calculation of the uplift, namely in such a way that the uplift was proportional to the difference between the volume porosity and the surface porosity. Although the structure of the formula was correct, one term was completely wrong, (the surface porosity must be replaced by unity) and, thus, it contradicted Archimedes' principle. In his treatise from 1913, he discovered another essential mechanical effect in liquid-saturated porous solids, namely the *effect of effective stresses*. This effect has led to one of the most important concepts in soil mechanics and in other fields where saturated porous solids are treated. This concept states that the stress state in such solids can be additively decomposed in a term governed by the porewater pressure and a term governed by the motion of the porous body. Fillunger (1913) stated: "It is less the actual individual stress problems than the complete agreement of the final results which lead straight to the conviction that a pressure-carrying liquid penetrating the masonry construction creates a pressure in the material equal in all directions."

He subsequently defined his statement more precisely regarding the causation of hydrostatic pressure by liquid in the pores of the masonry: "one can assume that the uniform internal pressure cannot cause a significant reduction of the strength."

It seems that Fillunger (1913) was the first author to state that the poreliquid pressure did not have any influence upon the strength of the porous solid. One year later, Fillunger (1914) criticized the so-called Lichterfelder experiments described by Rudeloff and Panzerbieter (1912), which showed a strong influence of the poreliquid pressure upon the strength of the porous solid material. Considering the discrepancy between the Lichterfelder test results and his theoretical investigations, Fillunger (1915) decided to carry out his own experiments concerning the strength of liquid saturated porous solids. He communicated the test results that he had gained from tests on unjacketed specimens of cement under different water pressures (100 and 200 atmospheres). His main test result was: "that the tension strength does not vary with the water pressure, ..."

Besides this comment that the tension strength did not change when the water pressure varied, he remarked that the porewater pressure did not affect the material behavior of the porous solid at all: "It is also shown that the water pressure penetrating into the pores is not able to create any explosive effect, as seems to have been assumed by some, but has been disputed by the author."

In 1914, Fillunger handed in some papers to the Technische Hochschule of Vienna in order to obtain the venia legendi, namely papers on the stress distribution in wedge-shaped bodies and, as a habilitation work, the already-mentioned treatise on the uplift problem from 1913. The expert Prof. Dr. Wighardt pointed out that Fillunger was correct to declare his paper on the uplift problem as his habilitation work because this treatise was the most important. Moreover Wighardt wrote: "... he (Fillunger) has enough of his own ideas, ... Occasional unskillfulness and small mistakes come certainly from the fact that the general mathematical training of the author did not occur on a broad basis." Furthermore, Wighardt remarked that the wide field of mechanics was open only to those persons who had had a solid mathematical training. Therefore, Fillunger only received the venia legendi for the narrower field of elasticity theory and strength of materials, and not for the intended field of mechanics. The colloquium took place on December 7, 1914, at 4 p.m., and referred to Airy's stress function, to some important questions regarding the strength of materials (in particular, to the elastic hysteresis), to the relation of this to the fading of metals, and to purely elastic phenomena. The habilitation committee voted unanimously to allow Fillunger to give a test presentation. Fillunger made three proposals: *Theorie der Festigkeit von Zughaken* (Theory of the strength of tension hooks), *Die ebenen Probleme der Elastizitäts-Theorie* (The plane problems of the theory of elasticity), and *Kurzer Abriß der Potentialtheorie und das Problem der Ausbreitung der Kräfte* (Brief survey of the potential theory and the problem of the propagation of forces). The committee chose the last mentioned title and, at 4 p.m. on January 15, 1915, Fillunger delivered his presentation in front of the members of the committee. He was well prepared and spoke clearly, fluently and with enthusiasm. The committee was satisfied and the habilitation procedure was concluded.

In the next years, Fillunger published several remarkable papers. In 1914 he continued his investigations on porous solids (Fillunger, 1914). He refined his uplift formula and, for the first time, investigated the friction phenomenon, which occurs during the flow of liquid through a saturated porous solid. In this paper, Fillunger pointed out: "Before I show how one can introduce the uplift into the calculation of barrages, we must still consider a second less known force, namely the capillary friction of the water." Then, Fillunger stated a formula for the friction force which turned out to be simple, though capturing the right effect.

However, he was not only occupied by porous media problems, but investigated also structural ones and as a result of his study of airplane problems he worked out papers on airplane-relevant mechanics in 1918 and even in 1928 (Fillunger, 1918 a, b, c, 1928 a). He published his valuable results in renowned journals.

Fillunger's reputation grew steadily. Thus, it was quite natural that he was offered a professorship at the Deutsche Technische Hochschule of Brno. However, he refused this offer and took over the aforementioned o. ö. professorship at the Technische Hochschule of Vienna. In the laudatory report, the search committee pointed out that Fillunger connected theory and experiment in an excellent way. In the time following, Fillunger was quite busy with teaching and with the organization of his chair. Thus, it was understandable that his scientific output in the form of papers was a little poor. In 1932, he was appointed as a dean.

Up to the end of the 1920s, the academic career of Professor P. Fillunger had taken a normal course, without any particularly noteworthy events. He was a well-known scientist in the field of mechanics, especially in porous media and elasticity theories, a strict teacher and an inexorable examiner. Moreover, in his neighborhood, he was known as a person possessing engaging manners and, within his circle of friends, he enjoyed decidedly great popularity. However, some colleagues reported that he always stood firm in his views concerning the fundamental principles of mechanics.

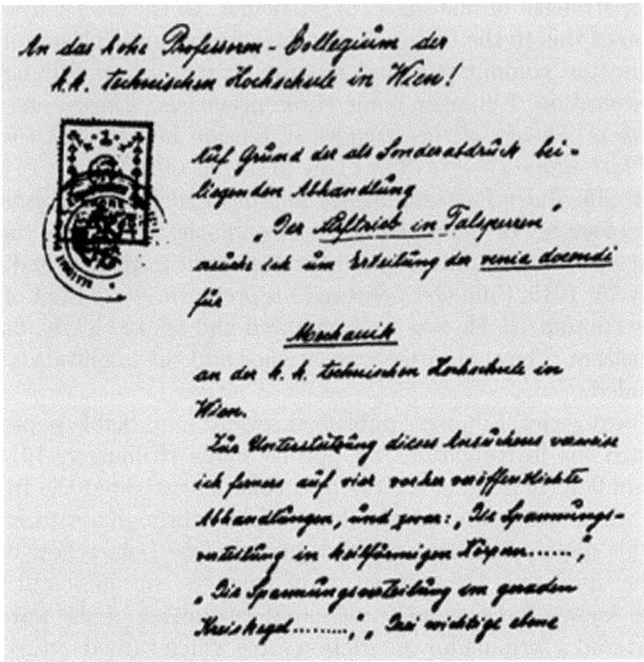

An das hohe Professoren-Collegium der
k.k. technischen Hochschule in Wien!

Auf Grund der als Sonderabdruck bei-
liegenden Abhandlung
„Der Auftrieb in Talsperren"
ersuche ich um Erteilung der venia docendi
für
Mechanik
an der k.k. technischen Hochschule in
Wien.
Zur Unterstützung dieses Ansuchens verweise
ich ferner auf vier weitere wissenschaftliche
Abhandlungen, und zwar: „Die Spannungs-
verteilung in keilförmigen Körpern......,"
„Die Spannungsverteilung im geraden
Kreiskegel.........", „Drei wichtige ebene

Fillunger's application for his habilitation.

Then, Fillunger changed his sociable behavior in one point, in a somewhat strange manner and without a perceptible cause. Suddenly, he felt called upon to act as a keen reviewer in all technical and mechanical questions.

In 1928, he published a polemical statement against the civil engineer Ottokar Stern (Fillunger, 1928 b), whose work on the foundation of concrete piles was strongly attacked. In the following period, the polemics continued. In his last statement on the aforementioned subject, he revealed his motivation: "An objective criticism is also in that case useful and necessary, if it pursues only that purpose to destroy a totally unfounded confidence in a scientifically indefensible theory. I take it upon myself to do it also in the future if it refers to a matter like this whereby excessive praising of a technical innovation could not only cause great material damage but, under certain circumstances, human life could also be jeopardized."

This statement is without a doubt the main strength in Fillunger's polemical criticism. However, this was in all cases, as we will see in the following events, connected with over-sensitive reactions to the criticism of his own work and statements. As Stern (1928) complained, the polemical discussion was caused by personal hostility which originated in a confrontation in a standards committee.

Three portraits of Paul Fillunger

In 1929 and 1930, a polemical discussion over the uplift problem took place between Fillunger and the Italian professor Hoffman, without, however, clarification of the problem. Hoffman (1929) accepted the structure of Fillunger's (1913, 1914) formula and praised his work on liquid-saturated porous solids. In contrast to Fillunger, Hoffman's idea was to change one term in Fillunger's formula slightly, which would yield the most unfavorable effect for the respective dam construction, thus increasing the stability of the building.

Fillunger was obviously deeply hurt by Hoffman's criticism. He answered this with heavy polemic because he was convinced that Hoffman's (1929) contribution was incomplete and, in the main parts, incorrect. Although the polemic did not lead to a clarification of the scientific issues, the dispute obviously encouraged Fillunger from then on to fight more strongly against errors and inadequate representations in scientific publications.

At that time it was not known that his strongest criticism, which would have devastating consequences for him, was soon to come out several years later. It was also not known in those days that he would become the founder of the theory of porous media with a scientific masterpiece.

His private life followed a quiet course. He married Margarete Gregoritsch, a young woman from Graz, on May 31, 1908. Their only son, Erwin Fillunger, was born on February 22, 1909. It seems that Paul and Margarete Fillunger lived in happy matrimony. Everyone who was closely related to the couple stated that there could not have been a more ideal marriage than that of Professor Fillunger and his wife. One of Fillunger's cousins pointed out that Mrs. Fillunger's husband was always her highest priority and that she always believed deeply in him. The partners were never separated, except during the time that Professor Fillunger was on duty during World War I.

The residents of the quiet, two-floor house with small front garden in the *Messerschmidtgasse im achtzehnten Bezirk* in Vienna with four apartments, in which the couple lived in a middle-class, simply furnished, four-room apartment for three years, spoke of the kind and discreet character of the couple.

However, several relatives reported that at home Fillunger was at times a *know-it-all*, as in his scientific life. Towards his son, he was correct and stern.

2.2 Karl von Terzaghi's Daily Life and Early Dispute with Paul Fillunger

In 1933, Karl von Terzaghi continued his consulting work in North-Africa, in Bou Hanifia (April and May). He had a small house built south of Oran; in the first floor he had installed the laboratory and the second floor served as apartment. He connected his business trips very often with private interests, vacationing and sight-seeing. Such was the case on this occasion. He visited Tunis and Carthage as well as Algier and Oran. And he undertook tourist tours with Ruth.

On June 24, they were on their way to Scandinavian countries, Sweden and Norway. They were very much impressed by Sweden's capital, Stockholm. The next stop was Oslo, capital of Norway. After visiting some points of interest in Oslo, they undertook a voyage to the north of Norway, the beautiful coastline with its marvellous fjords. The first station was Bergen, then Kirnna and Narvik.

In August, 1933, Karl and Ruth von Terzaghi visited England: New Castle, London, Oxford. They rushed further to Paris, where they met up with Ruth's mother. On August 23 they were back in Vienna and stayed there until October 30. During this time Karl worked on 5 articles for Engineering News-Record and a treatise on the Tower of Pisa. On September 15 they rode to visit Vera in Meran. She told them that she had joined the Nazi Party. Karl spoke with Vera's girlfriend about Olga. She told him that Olga exploited Vera. Moreover, Olga did not permit Vera to read at home in the evening – in order to save electricity.

Back in Vienna, Karl and Ruth admired a performance of Goethe's *Faust*. Gretchen was played by the famous Paula Wessely, Faust by Rehmann and Mephisto by Pallenberg[38]. They also enjoyed the marvellous balls in Vienna during the winter season.

Karl von Terzaghi spent his fiftieth birthday with some friends in a small boarding house in Radenthein, close to Spitall, Millstat and the Millstätter Lake. It was a happy evening with the showing of an Italian movie and with travel memories. Karl stated: "The climax of animal life. Now it goes physiologically downhill. However, a satisfactory occurrence, if one can look back on honestly achieved and recognized work." Unfortunately, on this day he was informed that Hofrat Professor Forchheimer, his patron and friend, had died in Vienna.

Back in Vienna, Professor Richard von Mises visited him at the end of October, 1933. Von Mises belonged to the first victims of Hitler's brutal and cynical policy. He was on his way to Constantinople.

Ruth and Karl continued attending ballet, opera and theater performances during the winter months. At that time, Vienna belonged among

[38] All three were well-known actors in Vienna and Paula Wessely, in addition, a famous movie actress before and after World War II.

the cultural capitals of the world. They saw the ballet opera *Das jüngste Gericht* and Tchaikovsky's Fantasy. Obviously, they liked the great actor Werner Kraus very much; within a short time they saw him twice, in Shakespeare's *Julius Caesar* and in Schiller's *Wallenstein (Piccolomini und Tod)*. In *Julius Caesar* the part of Brutus was played by Paul Hartmann, another great character actor. However, they enjoyed also some comedies, like *Dr.med. Hiob Praetorius* by Curt Götz, who also played the role of the doctor. Moreover, in the Josephstädter Theater they had a really funny New Year's Eve – and this, despite a dispute with Fillunger in the Café Kuhnhof in Vienna in the morning.

Karl spent the turn of the year at home with a bottle of Italian champagne under the lit Christmas tree and with Beethoven's music. "Last year: Hitler in Germany, cut-back of freedom in Austria, collapse of the economy in US, no perspectives in Sweden and England. Successes: observation of settlement, uplift in concrete ... death of Forchheimer on my birthday." At the beginning of 1934 the political situation in Vienna was getting worse. In Vienna there were demonstrations on the street and turmoil. On a Saturday evening Ruth ran into a crowd with policemen on horse backs. "Streets flooded with policemen and militia."

Karl von Terzaghi with assistants in Vienna (1933), from the left:
L. Rendulic, Mrs. Rendulic, W. Steinbrenner, Karl von Terzaghi,
Mrs. von Terzaghi, P. Siedek, K. Kienzl.

On January 31, the Professorenkollegium gathered for a meeting. The professors discussed the correct formulation of an official communication to the government concerning the arrest of eight students for about three hours, which had occurred the previous day, because of singing the *Deutschland-Lied* and firing off gun-salvos.

The political turmoil continued during the year of 1934. On February 12, a red revolt broke out, which the Christian-Socialist government put down with machine-gun and artillery fire. The Social-Democratic party was liquidated, Major Seitz was dismissed and replaced by the Christian-Social Minister, Schmitz. Karl saw that three policemen were shot to death. There were many other men who were killed. Professor Örley organized a collection for the relatives of the dead. On Friday, February 16, the situation had cleared. The road blocks were gone and the tram-traffic was back to normal; the weapons of the militia were collected.

On July 25, a National-Socialist revolt took place in several parts of Austria after the assassination of Federal Chancelor Dollfuß, which was put down by force of arms just as the red turmoil in February had been. Hundreds of men were killed in the combat, several members of the National-Socialist Party were executed, thousands of men were arrested. Some professors, as Wolf and Hartmann, sanctioned the actions of the government and insulted the National-Socialists.

At the end of March, 1934, Karl von Terzaghi considered giving up his academic career after the intended publication of a book on foundations and the second edition of his *Erdbaumechanik* – and returning to practice. "This under the impression of the discussions with Fillunger, political development in Austria and impressions of the study of the discussions of ... (two unknown persons, the author)(preparation of my last earth pressure article). Immense gap between theory and practice, becoming always larger. For example, earth pressure: professors and fellows write without a hunch about reality – 'l'art pour l'art' – and the engineers brood without a clue of theory. Kozeny, Zunker quarrel about nothing, hypothesis without any practical importance, inbreeding in scholarly circles, immoderate overestimation of handicraft achievements. As long as the academic profession ensures a sorrow-free existence and great freedom, it is worthwhile. However, the good times are gone. The 'foundation construction' answers all the questions which I had asked twenty years ago. Therefore the academic profession has fulfilled its purpose in my life and it is obvious to return to practice. I could leave Erdbaumechanik to Casagrande as inheritance." He repeated his wish to move to the United States in a letter to Arthur Casagrande from May 25, 1934.

As previously mentioned from 1932 to 1934 the dispute between von Terzaghi and Fillunger over the uplift problem occurred. Karl von Terzaghi was led to a new investigation of the uplift problem during the preparation for his main lecture in the fall of 1932, while studying Fillunger's papers from 1913 to 1930 on the uplift in dams. He came to the conclusion that it was impossible for him to share Fillunger's opinion. He visited Fillunger on November 21 and December 14, 1932, and told him of his doubts in order to be sure that he had not misunderstood his views. Since Fillunger could not dispel these, von Terzaghi had to prove the validity of his own views through experiments. These were performed during 1933 and yielded the results which von Terzaghi expected. In the second half of September, 1933, he showed Fillunger a manuscript in which he had described his results. On this occasion, von Terzaghi offered Fillunger a proposal that they should both present the central issue of the differences in their views in an oral discussion, and that they should publish the result of the investigations in an objective way, leaving any comment to the readers. In this way, the practical purpose of the dispute would be completely fulfilled without giving outsiders the sad spectacle of a polemic. Instead of agreeing to the proposal, Fillunger phoned Karl von Terzaghi on the morning of December 31, 1933, and asked him not to submit the manuscript for publication before he had talked with him. On this occasion, a meeting was arranged at 10:30 a.m. on the same day in the café Kuhnhof in Vienna. During this meeting, Fillunger ordered von Terzaghi, in a brusque way, to refrain from printing his manuscript. Karl von Terzaghi commented on this in his diary. "Two weeks spent with investigations of uplift, in order to be prepared for Fillunger ... Fillunger so pigheaded in his confused train of thought that he never will realize it. Urged me in all friendship to withdraw my paper on December 31, otherwise he has to tear me to pieces."

Paul Fillunger admitted this incident, however, he formulated a completely other version of von Terzaghi's statement in the questionnairy at the Technische Hochschule of Vienna in 1936/37.

Karl's struggle with Fillunger continued in January, 1934. "Three days of repulsive work with Fillunger discussion. An irrational quarrelsome person."

Despite this heavy dispute, von Terzaghi offered a further proposal in a letter from February, 1934, in order to avoid the impression of personal differences, at least in the public eye. However, Fillunger refused this proposal in a telephone call. Finally, the manuscript was published by von Terzaghi and Rendulic (1934).

The theoretical treatment of *capillary* forces in saturated porous media was closely connected with the investigations of the uplift problem and became also a part of the dispute between von Terzaghi and Fillunger, however, mainly on the part of Fillunger. Already in his first fundamental book on soil mechanics, von Terzaghi (1925a) had dealt with the capillarity problem. Later (von Terzaghi, 1933), he improved his investigations and concluded that with respect to the vaporization process at the down stream face of masonry dams, the suction for the poreliquid caused an additional pressure on the masonry: "... the surface stress of the water effective in the vaporization zone generates an additional pressure of the masonry in the region of the down stream

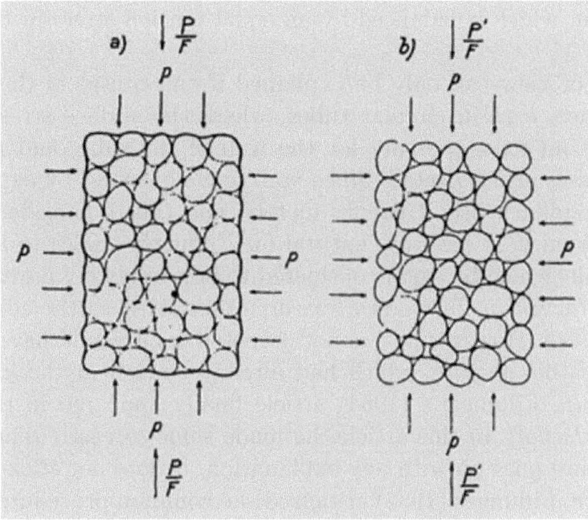

Investigation of the uplift problem
by von Terzaghi and Rendulic (1934)
(Determination of the effective surface porosity)

Einfluß des Wasserdruckes auf die Zugfestigkeit von Portlandzement-mörtel 1 : 3.

Überdruck in Atm. beim Zugversuch	Anzahl der Versuche	Zugfestigkeit in kg/cm²			Angewendete Zerreißvorrichtung
		Höchstwert	Niedrigster Wert	Mittelwert	
0	12	40·4	29·8	36·1	Michaelis-Wage
100	12	39·2	25·7	32·4	Apparat
200	12	36·5	26·8	31·6	Apparat

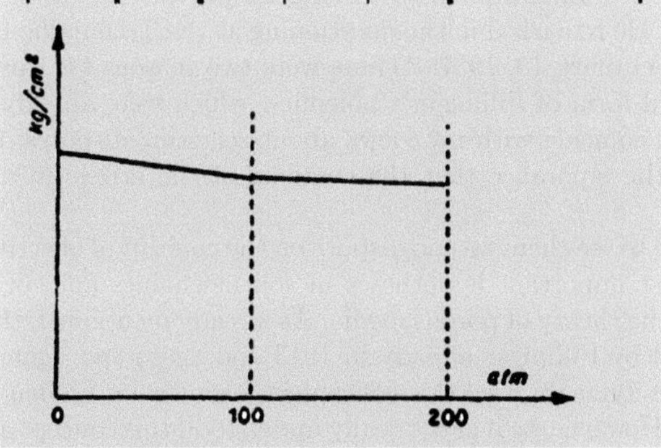

Fillunger's test results from 1915
concerning the effective stress principle

face of the dam, which is balanced by an equal tension stress in the flowing porewater."

This point of view can only be explained if one considers the profile of a liquid interface, e.g., in circular tubes, where the surface stresses of the meniscus cause an axial pressure for the wall of the tube, and radial and tangential stresses are neglected. Since von Terzaghi found by experiment a value for the capillary rise of twenty meters, and thus a considerable pressure on the masonry, it was only natural for Fillunger (1934) to attack von Terzaghi's result, which he again considered to be completely unrealistic. His article *Kapillardruck in Talsperren* was first submitted to the editor of the journal *Bautechnik*. However, it was not accepted, as it would have only continued the scientific dispute, which had already started in the journal *Die Wasserwirtschaft*. Fillunger's (1934) article finally appeared in the journal *Die Wasserwirtschaft*. In this article, he made some sarcastic remarks, e.g., "Although nature puts up with any explanation, it does not stick to them."

Furthermore, Fillunger (1934) criticized the common procedure for treating the capillarity problem: "The preceding considerations are not completely satisfying, since from the three principal normal stresses of the stress state, which are produced as a consequence of the surface stresses of the water in a saturated porous body, only one was calculated, namely the capillary pressure. In a pipe there are, beside the axial pressure, tangential and radial stresses. These are equivalent to certain stresses in a saturated porous body, however, only few reliable remarks can be stated." (It is amazing that Fillunger was not familiar with the valuable papers of his colleague Kozeny, 1924, 1927, on this subject.)

Although Fillunger (1934) strongly criticized the formula given by von Terzaghi (1933), he also maintained the main statement, i.e., that the capillary suction for the liquid caused an additional pressure for the solid phase.

Von Terzaghi didn't respond to Fillunger's attack concerning the capillarity problem. He remarked in the questioning at the Technische Hochschule of Vienna on February 10, 1937: "There were two reasons for this. The first lay in the awful form of Fillunger's polemics, which were already known to me and do not coincide with my views about scientific disputes; the second reason lay in the ignorance that the contents of the article in my opinion reveal."

It is strange to see there was no dispute on the concept of effective stresses, one of the most important hypotheses in soil mechanics and, of course, in other fields of the theory of porous media. As already mentioned, this concept was introduced by Fillunger already in 1913 and again and again improved upon. Karl von Terzaghi used this concept in a somewhat hidden way in his paper in 1923. However, he had not really understood this concept and did not recognize its meaning. During the preparation of the anniversary book: *From theory to practice in soil mechanics*, L. Bjerrum wrote to A. Casagrande at the end of the letter of May, 1958: "Finally I may mention that the explanation of Terzaghi's description of the development probably is due to the fact that he wished to emphasize that he already in 1923 recognized the effect of pore pressure on the shear strength of soil. However, he did not include this effect in Coulomb's equation in any of his early papers nor in Erdbaumechanik." It is obvious that the concept of effective stresses became clear in the first half of the 1930s through the extensive experimental and theoretical works of von Terzaghi's co-workers Hvorslev and Rendulic. In a paper by Rendulic (1935), there is no hint of effective stresses although Rendulic used the decomposition of the stress state for the mixture body with incompressible constituents (clay and water) in the correct way. One year later Rendulic (1936) defined the effective stress state as the state which is exclusively transferred in the solid porous phase and he formulated the concept of effective stresses: "The mechanical behavior of clay (deformations, strength etc.) depends only on the effective stresses of the clay." His formulation is similar to von Terzaghi's (1936) statement, as we will see in Section 3.1.

In 1933 and up to the middle of 1934, Karl von Terzaghi worked 5 months on the extension of his earth pressure theories. He published his results in a series of five articles in Engineering News-Records (see von Terzaghi, 1934). However, during his investigations, in particular the experimental ones, he had gained much more important results and he prepared some additional manuscripts in English; however, the editor rejected these because they were too detailed. Karl decided to edit these in a book in German and he quickly found the publishing house Julius Springer to realize the project. At that time, Karl von Terzaghi was occupied with some other book projects, on geological and seepage problems and, in particular, on the consolidation of clay layers with his friend O.K. Fröhlich. Maybe this was the reason that he engaged the civil engineer Dr. Alfred Scheidig at the beginning of 1934 as translator; Scheidig was a co-worker of Professor F. Kögler from the renowned soil mechanics laboratory[39] in Freiberg, Germany. Already in 1924 this laboratory had been founded by Professor Kögler and gained an excellent reputation in the following years. In 1938 they summarized (see Kögler and Scheidig, 1938) all their findings in a book with the title *Baugrund und Bauwerk* (Foundation Soil and Structure) which at the time was already appearing in its fifth edition. (Incidentally, in 1939 Professor Kögler committed suicide due to political reasons and health problems). Karl von Terzaghi knew Alfred Scheidig

[39] In the 1920s and 1930s some other well-known and recognized earthwork institutes (laboratories) were founded in Germany; *Preußische Versuchsanstalt für Wasserbau und Schiffbau* (Prussian Research Institute for Hydraulic Engineering and Ship Construction) in 1904 with the earthwork department, founded in 1927. At this institute such famous engineers and scientists were working as Professor Krey and Ehrenberg; *Hannoversche Versuchsanstalt für Grundbau und Wasserbau* (1927) at the Technische Hochschule Hannover, where also already earth-statical investigations had been performed (Professor O. Franzius and Dr. Streck). Since 1932 Dozent Streck had been lecturing on soil mechanics in Hannover; *Deutsche Forschungsgesellschaft für Bodenmechanik (Degebo)* (German Research Association for Soil Mechanics) founded in 1928 (Professor Hertwig).

very well, for in 1931 Scheidig visited the earthwork laboratory in Vienna and worked there. After his return to Freiberg, Karl von Terzaghi supported him in every respect. Scheidig, in turn, dedicated his monographs on loess to von Terzaghi "in honor and gratitude." Karl wanted Scheidig to deliver a liberal, press-ready manuscript. However, Dr. Scheidig sent in only a word-for-word translation. Karl was upset about Scheidig's translation work. He wrote to him: "I have invested altogether two years of hard work into the series of articles and have disclaimed the payment by Engineering News-Records, in order to obtain means for covering the salaries of scientific auxiliary assistants, in place of an honorarium." Scheidig was not able to fulfil Karl von Terzaghi's wish because he had only little knowledge of the English language and had not understood the factual contents of Karl's manuscript. After letters, exchanged over some months, the relationship ended with the result that the book was never published. Karl von Terzaghi later used parts of his manuscript in his book *Theoretical Soil Mechanics*.

The motives for Karl's wish to obtain a liberal translation were not quite clear. It is known that it is not easy to catch the *spirit* of a new treatise if one is not familiar with the subject to be translated. Why did he ask Dr. Scheidig from the renowned Freiberger laboratory for a translation and not Arthur Casagrande?

From July 12 till the 30, Karl spent his vacation together with Ruth in Hintertux, where they had almost nothing but rain. Therefore they moved on to St. Moritz, Lugano, and Meran.

Karl and Ruth von Terzaghi in Tyrol, 1934

Karl and Ruth von Terzaghi studying a map, 1934

Karl and Ruth von Terzaghi in the beloved Alps, 1934

Karl and Ruth von Terzaghi on vacation, 1934

Karl von Terzaghi at Hintertux, 1934

Fritz Todt,
Hitler's Generalinpector for
German Road construction

On September 21, he began work on a book about foundation construction and in December, Karl gave several talks in Paris, London and The Hague.

On May 14, 1935, Karl and Ruth rushed for an extended trip to Turkistan, where Karl had to work out his expert opinion of a large earth dam. It was an impressive tour for both and Karl as well as Ruth reported in two long articles about this adventurous trip. At the end of June, they were back in Vienna.

In the fall of 1935, Karl von Terzaghi told about his contribution to designing the foundation of the Congress hall in Nuremberg. "On October 27, telephone call from Berlin. Todt[40]: Reichskanzler wishes expert opinion foundation of Reichspartei Congress hall. As soon as possible. Promised departure to Nuremberg Thursday evening. Largest construction since Cheops pyramids

[40] Fritz Todt, born in Pforzheim on September 4, 1891, studied civil engineering at the Technischen Hochschulen Munich and Karlsruhe, after finishing the elementary and high school.

He worked in the beginning on hydroelectric power plants and later on road construction, in particular, on the highway. In 1922, he entered the Nationalsozialistische Deutsche Arbeitspartei (Nazi party). Fritz Todt did his doctor thesis in 1932 and became appointed Generalinspektor für das deutsche Straßenwesen (Generalinspector for German road construction) after Hitler's assumption of power, answering directly to Hitler. He was responsible for the total construction industry, in particular, for the construction of the Reichsautobahnen (freeways). Todt founded the *Organisation Todt*, which built the Westwall (Western Wall) and later shelters for U-boats, concrete dugouts and roads in the occupied countries – and also labor camps. In 1940 he became minister of *Bewaffnung und Munition* (of armaments and munitions). Fritz Todt died in an airplane crash in 1942. Albert Speer became his succesor.

... Orderer Administration union Reichparteitag Nuremberg, Mayor Liebel." Karl started on October 31, at 10:30 p.m. to Nuremberg and arrived there on November 1, at 8:00 a.m. Leo Casagrande and Ernst Gottstein, picked him up at the railway station. At 9:00 a.m. a meeting, chaired by town councillor Brugmann, was held in the town hall. On Saturday, Karl studied again the documents and in the late afternoon he headed with an express train to Munich where he stayed in the Exzelsior hotel.

Fritz Todt announcing "to his Führer"
the completion of the first route of the Reichsautobahn
Frankfurt on the Main – Darmstadt on May 19, 1935

3. Intermezzo

3.1 Karl von Terzaghi's Career Peaks

In the middle of the 1930s Karl von Terzaghi's career reached several peaks. The first one was a strange and unbelievable meeting with the Führer, Adolf Hitler, discussing the foundation of the Congress Hall in Nuremberg (see Section 2.2).

In the morning of Sunday, November 3, 1935, Karl von Terzaghi met the General Inspector (for German road construction), Dr. Todt and attended the opening ceremony of the Ludwigsbridge. After the ceremony he rushed by train to Garmisch-Patenkirchen. During a break in a tunnel a small police man with a pointed beard rushed in. "Telephone message from Todt from Munich. Führer wishes to speak to me in his private apartment at 5 p.m. ... Lunch in friendly Bavarian pub." After lunch he rode back to Munich, to Todt's apartment where he had tea from 5:00 to 7:00 p.m. "To the Führer by car, around 10 minutes. In quiet street with calm dark-brown houses. In front of the house long row of cars. Two huge guards, dark death's head uniforms, raised hand for Hitler greeting. Stairs upwards, reception hall. Large, round table, above the chimney Stuck's[41] Sünde. Many coats in the closet, cigarettes on the table. Almost one minute, Führer appears, young and elastic, totally in brown; merry, friendly face. Apologizes for modest home, leads into the reception hall with long, narrow table. Longitudinal walls empty, on each narrow side a painting by Feuerbach.

At first reported about congress hall." Then Karl von Terzaghi made a joke and Hitler laughed heartily. "Will slowly develop foundation. Then, I mentioned ugly granite. Now, he became lively. The marvellous collection of useful stones in Vienna museums! He will establish a similar one in Munich. Searching for the place of production. Describes with enthusiasm and pointing to beautiful stone, which once he saw. Has the intention of livening up stone-cutter business. Twenty million marks, distributed over five years. Lively face, highly domed forehead, strong projection above nose. Hair falls uncombed

[41] Franz Stuck, born in Bavaria on February 23, 1863, was a painter, sculptor and architect and one of the first pioneers of Art Nouveau. He painted, influenced by Böcklin, pictures with symbolic-allegoric themes, e.g. Die Sünde (Sin). Stuck died on August 30, 1928.

over half of the forehead. Only the weak, feminine mouth and the small ugly moustache above it are disturbing.

Then to the convention: A splendid family festivity – gathering of old friends and comrades in arms. You have to participate in that! An American came sceptically and soberly; now he lectures on propaganda in USA. Realized how we are in the right ...

Congress hall should last like a Roman monument. Human beings should think in eternities. Manoevre field city tournament field. A fortune that joy in the use of weapons runs in man's blood. Come from far away in order to take a look ... Will see what we will make out of Munich! Whole street block will be diverted, replaced with splendid buildings.

Conversation lasted three quarters of an hour. Accompanied me to anteroom, introduced me to Mr. and Mrs. Göbbels, 'his friends'. Göbbels small, slender, long head with curly, black hair, head inclined at the shoulder and grinning in the friendly way, Mrs. Göbbels made up and powdered, impression of lady from theater. Behind the pair two giants in dark uniforms, physician in ordinary and adjutant. Göbbels spoke about bad foundation proportions, mud stream, in Berlin. 'Will probably also need you there urgently'. Farewell from Führer with hearty handshake and thanks that I came to Nuremberg.

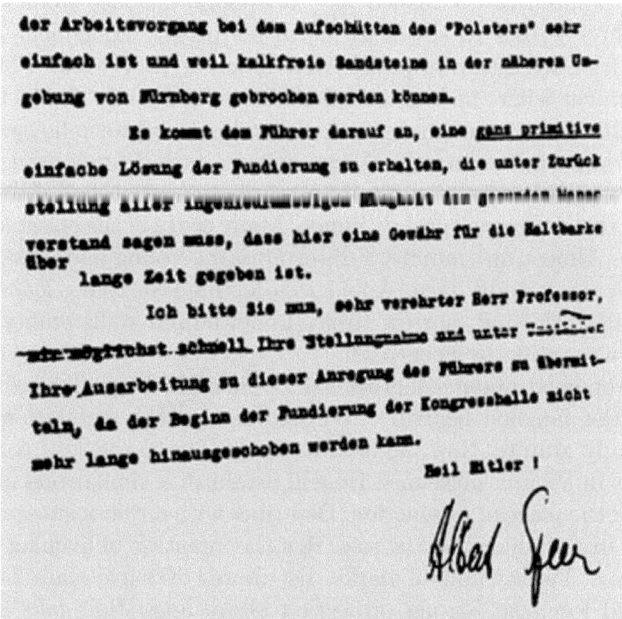
der Arbeitsvorgang bei dem Aufschütten des "Polsters" sehr einfach ist und weil kalkfreie Sandsteine in der näheren Umgebung von Nürnberg gebrochen werden können.

Es kommt dem Führer darauf an, eine ganz primitive einfache Lösung der Fundierung zu erhalten, die unter Zurückstellung aller ingenieurmässigen Klugheit [illegible] verstand sagen muss, dass hier eine Gewähr für die Haltbarkeit über lange Zeit gegeben ist.

Ich bitte Sie nun, sehr verehrter Herr Professor, mir möglichst schnell Ihre Stellungnahme und unter [illegible] Ihre Ausarbeitung zu dieser Anregung des Führers zu übermitteln, da der Beginn der Fundierung der Kongresshalle nicht mehr lange hinausgeschoben werden kann.

Heil Hitler !

Albert Speer

Letter of Hitler's architect Albert Speer from the beginning of February, 1936, to Karl von Terzaghi concerning the foundation of the Congress Hall

Impression of this lively, imaginative personality was strong and lasting. T. (Todt, the author): He speaks about every subject and with the same expert knowledge. Nearly inexplicable. Fantastic memory ...

Dinner with Leo (Casagrande) in the Würthenberger Hof. Envyed me genuinely for the meeting."

On Monday, November 4, 1935, Karl von Terzaghi was back in Vienna. "Ten days strenuous work on the expert opinion for the congress hall. They had made a bad mistake. Forgot to point out diffusion phenomena through foundation rings."

There is no evidence for this amazing and bizarre story. It is hard to believe that Hitler received Karl von Terzaghi at all. Hitler's partner for all questions concerning buildings and construction problems was Albert Speer, his architect. And thus, it was quite natural that Speer had written a letter to Karl von Terzaghi on February 1, 1936, wherein he explained the problems and wherein he asked von Terzaghi to send him his comment on Hitler's idea.

There is neither any indication of the meeting with Hitler in Speer's letter nor in any other documents in the file on the foundation of the congress hall (in the Terzaghi Library). Also in a letter written by von Terzaghi's former assistant Rendulic, in which he admired Hitler's clear recognition of the foundation problem of the congress hall, one cannot find any remark about the meeting with Hitler.

It is also hard to believe that a stranger (for Hitler) could come, without having been screened, into Hitler's apartment and could spend forty-five minutes alone with the Führer. Hitler was Reichskanzler – and dictator; he was always guarded.

Moreover, Karl von Terzaghi kept not only the extended diaries, discovered at his home in 1995/97, which are the basis of this treatise, but also an incomplete set of diaries with short entries which have already been known for a longer time and which are preserved in the Terzaghi Library in Oslo. In these diaries, nearly every day is mentioned. However, the entries for 1935 end in the middle of October.

The main objection to Karl von Terzaghi's version of the meeting with Hitler comes from the diary entry of Göbbels from Sunday, November 3. Göbbels gives a clear overview of his schedule for the day. He wrote on Tuesday, November 5, about Sunday, November 3:

"November 5, 1935 (Tue)
Sunday: Inauguration of new Ludwigsbridge.
Fiehler[42] speaks at length, Führer good. Beautiful bridge. **Much praised**!
Brown house: new party buildings, topping-out ceremony, a marvellous construction, masterpiece of Troost[43]. Emotional moment to stand in front of these new buildings. Schwarz[44] speaks, full of justified pride. Then, a witty

[42] Karl Fiehler was Mayor of Munich from 1933 till 1945.
[43] Professor Troost, Hitler's previous architect.
[44] Franz Xaver Schwarz, treasurer of the Nazi Party.

foreman has six glasses of wine. There the buildings now stand. Monuments to our activities. The Führer is totally happy. Visit of the Führer buildings ... In the morning: botanic garden ... Lunch in the Löwenbräu. With the workers. Führer speaks very heartily ... Later we are with him for coffee. He is glad and very kind to Magda and me. Gabbed a lot. Then farewell. Hotel still a little reading. And then back to Berlin. To the beloved children. Early arrival. Grey fall ... in Berlin. Monday. Immediately to the children. Monday: Back to work."

In no place of his entry from November 3, does Göbbels mention Karl von Terzaghi. Maybe von Terzaghi was too unimportant for him. Unfortunately, Göbbels did not note the exact time when he left Hitler. However, Göbbels and his wife had coffee with him which usually happens in Germany around three or four o'clock in the afternoon. It is not very likely that Göbbels and his wife stayed in Hitler's apartament so long, till around eight o'clock, when Karl von Terzaghi claimed to have left. Moreover, further research has also not revealed any hint of the supposed meeting[45].

Karl von Terzaghi had also told his students and his relatives that he had met Hitler several times, once on the Obersalzberg. However, there is no evidence in the literature or in documents so far. There is also no hint to Karl von Terzaghi's appointment in the files of the Parteikanzlei (office of the party), of Martin Bormann's calendar, and in the unpublished part of Göbbel's diaries.

After the alleged reception by the Führer he prepared for the next highlight in his career. The Civil Engineering Department of the Technische Hochschule, Berlin, had invited him to lecture on soil mechanics during the winter semester 1935/36. On Thursday, November 14, 1935, he rode to Graz in order to say good-bye to his relatives and on Saturday he rushed together with his wife to Berlin. His assistants, e.g. Hvorslev and Fröhlich were at the Vienna railway station to see him off. On Saturday Karl and Ruth met Leo Casagrande, Rendulic and Gottstein with their wives. The main subject of their first talks was his former assistant Loos, who had, obviously, plotted against Karl von Terzaghi.

On Monday, November 18, 1935, at 4:30 p.m. Karl von Terzaghi opened his guest lecture series at the Technische Hochschule in Berlin-Charlottenburg with a talk on the practical application of soil mechanics in front of a crowd of around 300 persons. He was warmly welcomed by the head of the Civil Engineering Department, Professor Agatz, who also thanked the General Inspector for German road construction, Dr. Todt, for his appearance. The lecture was well received and frequently interrupted by applause.

Nearly every day, the von Terzaghis were invited to lunch and dinner by his former assistants, by Dr. Todt, by colleagues, friends and acquaintances –

[45] Karl von Terzaghi disclosed a large amount of detailed knowledge in context with his reception by Hitler. There are two possibilities for this. Either he was really received by the Führer or he was informed by a person (Todt?) out of the inner circle of the Nazis.

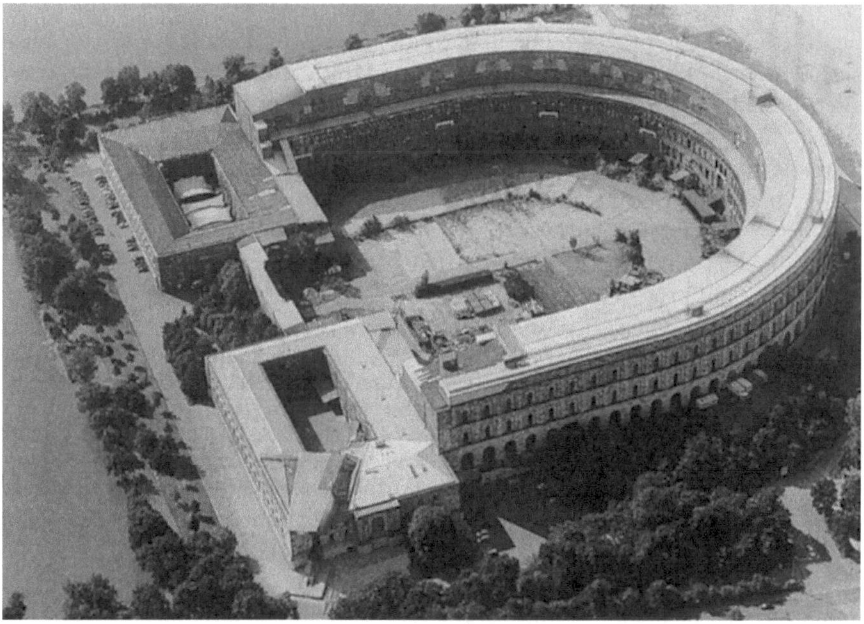

Torso of the Congress Hall in Nuremberg

and Karl spent a lot of time solving revenue problems. However, much more enjoyable were the meetings and cocktail parties in the exclusive hotel Adlon and visits to theaters; e.g. in December Karl and Ruth attended a splendid performance of Friedrich Hebbel's *Gyges und sein Ring* with their favorite actor Werner Kraus as Gyges.

On Thursday, December 15, Karl rushed again for Nuremberg and arrived at 4:17 p.m. Immediately he went to the town hall, to the city councilor Dr. Brugemann. The commission was scheduled to meet in the late afternoon. Dr. Brugemann reported that the *Führer* had visited Nuremberg again and had stated that he would prefer that the congress hall be founded on rock. This was understandable because Hitler wanted to build the building to last at least a thousand years.

Gradually the members of the commission arrived: architect Ruff[46], government surveyor Erdmannsdorfer, two engineers and Professor Otto Graf from the material testing board of the Technische Hochschule of Stuttgart.

Professor Otto Graf suggested an isolated base slab. The isolation was necessary because of the aggressive ground water. Professor Graf claimed that he could guarantee the durability of the isolation and, thus, the discussion of the committee members was finished. However, some days later Karl von Terzaghi doubted the solution. In a letter dated December 21,1935, to

[46] Franz Ruff had designed the congress hall together with his father Ludwig.

Brugmann, he stated: "On Wednesday I began to consider quietly the technical details of the process. The first result was my express letter of the 17th of this month. Then, I remind myself of the difficulties with which the contractors had to fight concerning the isolation during the construction of the subway in New York, where I myself was active as a consultant." Moreover, he made inquiries about the durability of the isolation with the proposed copper plates in Berlin. He got useful support from Dr. Leo Rendulic, his former assistant. The result was that experts estimated the life-span durability at about a maximum of a hundred years, if the work would be flawlessly executed. Thus, the relatively short time of the durability was one reason that Hitler rejected Graf's design. The other was that there was no guaranty that the isolation would not be damaged at least at one place so that the aggressive ground water could enter forcibly into the foundation body and destroy the concrete. Hitler's rejection was imparted to Karl von Terzaghi by Dr. Todt in a letter of Leo Casagrande from January 28, 1936, and in a letter of Albert Speer around February 1, 1936. Leo Casagrande wrote: "The *Führer* has rejected the deep foundation of the congress hall, because he is of the opinion that, if the isolation is violated only on one place, the whole foundation will be lost in the course of time ... Mr. Dr. Todt is glad about your victory, which came about completely without the influence through his person, only through the fantastically correct thinking of the *Führer*" and Leo Rendulic remarked in a letter to von Terzaghi of February 5, 1936: "the whole excavation is to be filled with granite stone in mortar. As far as I remember you have already expressed such an idea. I admire thereby only the clear recognition of the *Führer*, who made the decision without knowledge of your objections against the isolation proposal. It is rumored that Dr. Todt is happy that Prof. Graf got a small damper."

Finally, a very simple solution of the *Zweckverband Reichsparteitag Nürnberg* was applied. The bad soil was removed and replaced by pure sand which was compacted. On this compacted soil, the foundation was constructed. Karl von Terzaghi accepted the solution; for his efforts he received five thousand marks, a remarkable fee if one considers that the Volkswagen Beetle cost under one thousand marks at that time. Later, he claimed in his diary that the performed solution of the foundation by the Zweckverband was his own proposal.

His overestimation of his own capacities culminated some weeks later when he compared his abilities with those of Dr. Todt. "Todt is a fully unscientific man. Does not understand the consequences of my ideas. Wants to become leader of German technology. In reality the complete technological society behind me, Todt has to recognize this. Plan organizing all institutes of Europe (meant are the Material Testing Offices, the author)." Besides, Karl had expressed his outstanding abilities (in his opinion) already in a diary entry of February, 1921: "If smarter people had been active in civil engineering, no white spot would be left on the map already for a long time. Now, the white spot has vanished and I have contributed the lion's share."

On December 22, 1935, the von Terzaghis were on their way to Hamburg to spend the Christmas holidays in the northern part of Germany. From Hamburg they rode by train to Goslar, a small lovely city in the mountains of the Harz. The next stations were Hildesheim and Hanover, where they visited Professor Franzius, who was involved in clarifying a bad accident at the subway site in Berlin. Karl had discussions with Dr. Streck, Franzius' co-worker, who installed an effective soil mechanics laboratory at the Technische Hochschule of Hanover.

Back in Berlin Karl von Terzaghi visited the State Material Testing Office and he was impressed by the variety of experiments and by the equipment in the departments. In the evenings he had the leisure to take part in cultural events. For example, he saw Leni Riefenstahl's *Tag der Freiheit – Unsere Wehrmacht 1935*. "Imposing spectacle by Leni Riefenstahl, bursting with great talent, back light, jump from individual into the mass. Bounded mass of tanks, long rows of conical guns, around eight or ten meters long. Aviator wings and aviator defence, cavalry, songs and combat scenes in quick cuts, an iron era. And drone speech of the Führer to his soldiers. They do not have to find the glory of the army but only to keep it."

On Monday evening, January 13, 1936, Karl von Terzaghi gave his first talk of the final series in front of around 150 engineers; most of them came from building administration and industry. Dr. Todt opened the evening with a warm speech. He remarked that he had never met an engineer who combined science and practice to such a marvellous extent as Karl von Terzaghi. He presented a silver cigarette-case with the road system of Germany in memory of Karl's time in Berlin.

On Wednesday, January 15, Karl von Terzaghi delivered his last lecture. "Brief, warm speech of Agatz ... Strong applause at the end of the talk."

In the evening, he and Ruth were on their way to Bremerhaven via Hamburg in order to embark for the United States. Karl was invited to lecture at Harvard University during the spring term and to act as president of the First International Conference on Soil Mechanics and Foundation Engineering at Harvard in June, 1936. During the voyage, he began to work out an opinion paper regarding the failure accident at the subway site in Berlin. After landing in the United States, he continued his work on the paper, first in New York and then in Boston. Besides, he met several friends in New York, had lunch or dinner with some important persons and prepared his lectures for Harvard.

At the beginning of February he moved to a small, nice office in the first floor of the Pierce Hall, Department of Civil Engineering at Harvard University. Karl and Ruth occupied an apartment close to the Germanic museum. On Wednesday, February 5, 1936, Karl von Terzaghi gave his first talk in the Pierce Hall, introduced in a friendly tone by Arthur Casagrande and Dean

Clifford, in front of 120 listeners, 25 students and the rest, engineers from Boston and the surroundings. Besides his lecturing he was deeply involved in scientific work, e.g., at the end of February and the beginning of March he worked restlessly on the capillarity problem, obviously with little success, since there is no essential contribution from Karl von Terzaghi known in the literature.

During cocktail parties, lunch or dinner, the situation in Germany played an increasing role in the conversations, as well as at a party on March 10. The party was marked by great excitement about the German occupation of the *Rheinland.* Hitler had broken in a particulary hazardous stroke not only the Treaty of Versailles but also the Locarno Treaty voluntarily concluded by Germany in 1925. It appeared for a moment that France would answer with mobilization and counter the move by marching in. Some Reichswehr generals were afraid of that and had warned against occupying the Rheinland. Hitler, however, believed that France would not act – and he was right. This feeling was possibly his greatest internal political success. From that point on he was sure that he could also complete his international plans, like the annexation of Austria and parts of Czechoslovakia.

On March 28, 1936, Karl von Terzaghi rushed for the West in order to lecture and to visit construction sites and labs in Chicago, Urbana, Indianapolis and Washington D.C. Moreover, he resumed old relations to the Massachusetts Institute of Technology (M.I.T.) again in April, 1936.

On May 9, Karl received the information that the foundation of the congress hall in Nuremberg would be performed as proposed by the commission in Nuremberg. In his diary he remarked: "Notice that my Nuremberg project will be carried out."

One month later, Karl and Ruth moved to Winchester into a grey house with high gables hidden between tree-tops. "Every piece in the house, from the engravings on the wall to the ash-tray and the table lamps, were chosen with love."

On June 19, the First International Conference on Soil Mechanics and Foundation Engineering, which had been organized by Arthur Casagrande, started at the Rockefeller Center in New York City. von Terzaghi, as the president of this conference, welcomed members from nearly twenty different countries, with the following address: "The opening of this Conference is an event of unusual significance. It represents the first international council in the perpetual war of the civil engineer against the treacherous forces of nature concealed in the earth. Due to scattered and world-wide efforts extending over a period of 25 years, new and efficient weapons have been forged and the prime purpose of our meeting consists in discussing the means of exploiting the advantages thus secured. For the sake of brevity these recent developments have been given the name of soil mechanics. The transition from the classical theories of the pre-war generation to soil mechanics is synonomous with a transition from a purely abstract treatment of the problems of soil behavior to

Karl von Terzaghi (46) and Arthur Casagrande (76) at the conference in Harvard

one based on an intimate knowledge of the manifold and complex properties of the different types of earth. The validity of the older theories of earth pressure and earth resistance was limited to ideal materials whose properties can be described in five lines. However, in order to describe the practically important properties of earth such as nature has produced, one needs a good-sized book. As a consequence, the older theories failed in a great number of cases of outstanding practical importance. This, in a nutshell, was the reason for the necessity of a radical departure from past practice ...

Origin of Soil Mechanics. Ten years ago the investigations which led to this Conference still had the character of a professional adventure with rather uncertain prospects for success. This adventure began a short time before the war, simultaneously in the USA, in Sweden, and in Germany. It was forced upon us by the rapid widening of the gap between the requirements of canal and foundation design and our inadequate mental grasp of the essentials involved ...

I myself, prior to 1912, worked as a superintendent of construction. Year after year, in the Austrian Alps, in Transsylvania, and in Russia, I had ample opportunity to witness the striking contrast between what we expected when digging into the earth or loading it, and what really happened. Deeply impressed by the fundamental futility of pertinent theoretical knowledge, I came to the United States and hoped to discover the philosopher's stone by accumulating and coordinating geological information in the construction camps of the US Reclamation Service. It took me two years of strenuous work to discover that geological information must be supplemented by numerical data which can only be obtained by physical tests carried out in a laboratory. The observations which I made during these years crystallized into a program for physical soil investigations which looked as if it could easily be carried out in one year. In reality, the research activities extended over a period of eight years.

Period of Transition. All these early efforts which were started before the war and carried on by isolated groups or by individuals had one important feature in common. They were still guided by the intention to establish a science of soil behavior comparable to the science of bridge design. The major part of the college training of civil engineers consists in the absorbtion of the laws and rules which apply to relatively simple and well-defined materials, such as steel or concrete. This type of education breeds the illusion that everything connected with engineering should and can be computed on the basis of a priori assumptions. As a consequence, engineers imagined that the future science of foundations would consist in carrying out the following program: Drill a hole into the ground. Send the soil samples obtained from the hole through a laboratory with standardized apparatus served by conscientious human automatons. Collect the figures, introduce them into the equations, and compute the result. Since the thinking was already done by the man who derived the equation, the brains are merely required to secure the contract and to invest the money. The last remnants of this period of unwarranted optimism are still found in attempts to prescribe simple formulas for computing the settlement of buildings or of the safety factor of dams against piping. No such formulas can possibly be obtained except by ignoring a considerable number of vital factors.

Unfortunately, soils are made by nature and not by man, and the products of nature are always complex. After a decade of mental and physical experimentation in the newly developed field, it became obvious that the method of approach must be radically changed. The design of bridges and of other purely artificial structures requires only a knowledge of mechanics. Theory governs the field and experience is a matter of secondary importance except for that acquired over the drafting board. The theoretical results can be depended upon, because the equations contain no important element of uncertainty. However, as soon as we pass from steel and concrete to earth, the omnipotence of theory ceases to exit. In the first place, the earth in its natural stage is never uniform. Second, its properties are too complicated for rigorous theoretical treatment. Finally, even an approximate mathematical solution of some of the most common problems is extremely difficult. Owing to these three factors, the possibilities for successful mathematical treatment of problems involving soils are very limited. In bridge design, the theory provides us with certainties and eliminates the necessity for observations on fully-sized structures. In soil mechanics, the accuracy of computed results never exceeds that of a crude estimate, and the principal function of theory consists in teaching us what and how to observe in the field. Whenever we explore the natural soil by drilling a hole or by extracting a sample, we alter its state even before the direct contact between the soil and the tool is established, and the effect of this change on the results of our tests can only be learned by experience. The theories which we apply in order to make the step from the test results to a numerical estimate of the effect of our engineering operations are bound to be based on radically simplified assumptions. The importance of the difference between theory and reality can again be learned only by experience. It depends on the type of soil to a large extent. The Proceedings of this Conference contain a great number of instructive examples. Finally, a natural soil is never homogenous. Its properties change from point to point, while our knowledge of these properties is limited to those few spots at which the samples have been collected. To get information on the importance of the error produced by our inadequate knowledge of the deposits, we are compelled to compare the results of our forecast to those of direct measurement in a great number of cases. Owing to these facts, successful work in soil mechanics and foundations engineering requires not only a thorough grounding in theory combined with an open eye for the possible sources of error, but also an amount of observation and of measurement in the field far in excess of anything attempted by the preceding generations of engineers. Hence the center of gravity of research has shifted from the study and the laboratory into the construction camp where it will remain. The first fruits of this revised and essentially empirical attitude towards the problems of earthwork engineering are assembled in the Proceedings of our Conference ...

The Conflict between Theory and Reality. One of the outstanding impressions which I got while preparing the abstracts of pre-war publications was that of a steady decline of the capacity for careful observation after the eighteen-eighties. Prior to about 1880, a surprisingly great number of stimulating field observations were published by engineers ... At the time when the theories originated, their authors were still keenly aware of the bold approximations involved, and nobody thought of accepting them at face value. As the years passed by, these theories were incorporated into the stock of knowledge to be imparted to students during the years of their college training, whereupon they assumed the character of a gospel. Once a theory appears on the question sheet of a college examination, it turns into something to be feared and believed, and many of the engineers who were benefited by a college education applied the theories without even suspecting the narrow limits of their validity. If the structures designed on the basis of these sacred theories stood up, their behavior was considered to be normal and not worth mentioning. If they failed, it was an act of God, which should be concealed from the eyes of mortals, who might believe that the designer was poorly grounded in theory. This uncritical attitude towards the problems of earth behavior induced a growing resentment of those who had eyes to see against the theoretical textbook wisdom ...

However, the feeling of resentment against unwarranted generalization does not suffice to transform an accumulation of haphazard professional experience into a store of knowledge and of general usefulness. In order to accomplish such a transformation, three conditions must be satisfied. First of all, there must be a generally accepted method for describing the soils to which the individual experiences refer. Conventional terms such as 'fine, water-bearing sand' may mean almost anything between a loose accumulation of small grains, incapable of sustaining an appreciable load, and a stratum which is almost as hard as rock. The terminology must be based on numerical values of some soil. Otherwise it is worthless. Second, the observation methods must be reliable; otherwise there is too wide a margin for interpretation. If an observer claims that a building did not show any signs of settlement, the structure may have settled through a distance of one-tenth of an inch to four inches, provided the settlement was uniform and the distance to the neighboring structure was appreciable. Finally, the report on the observation must be accompanied by a statement of all the vital factors which were likely to have influenced the object of the observation. Otherwise the observation cannot be used as a basis for a valid conclusion. In order to satisfy this third requirement, the observer must be familiar with the physics and mechanics of the observed phenomenon. Thus, for instance, no valid conclusion can be derived from the results of a settlement observation on a building covering an area of 100 by 100 feet unless we have at least reliable geological information concerning the nature of the subsoil to a depth of at least 150 feet. In one of the cases which I had under observation, a building settled more than one foot owing to the compression of a layer of clay located between a depth of 100 and 130 feet below the surface of the ground ...

Truth and Fiction in Textbook engineering ... Every theory consists of three parts, a set of assumptions, a process of reasoning, and a final result. Since the validity of the reasoning can easily be verified, it suffices to concentrate our attention on the first and last parts. Each of these may be dissolved into words expressed by symbols and figures. The first requirement for an acceptable theory should be that the words have a definite meaning. Many of the terms which are used in textbooks on foundation engineering have a very vague one, if any. In this connection, the term 'safe bearing value of piles' may be mentioned. Some eight years ago, a very expensive factory was established on a whole forest of piles, between 60 and 80 feet long. The machinery erected in this factor was extremely sensitive to unequal settlement. The bearing capacity of the individual piles was most satisfactory. According to all the textbooks and manuals relating to this subject, the load on the piles was equal to or smaller than one-half of the safe bearing value. Yet the owner of the factory refused to share this opinion, because some parts of his factory settled through a distance of one foot. In western Austria stands a post-office building with continuous footings on a very compact bed of sand and gravel, 23 feet thick. The building exerts a pressure of 2.5 tons per square foot on the ground. I do not know of any building code or of any textbook which does not contain a very much higher figure for the safe bearing value of such a stratum. Nevertheless, the settlement of the building ranged between two and three feet. The same books which inform the patient reader on the safe bearing values also contain instructive tables with the values of the coefficient of internal friction of fat and of lean clays and loams. Yet with some skill and experience in laboratory procedure, one can get almost any specified friction value for a given clay. A score of other examples could easily be added.

Considering these unpleasant facts, one of the first requirements for a clean-up in the field of foundation engineering is insistence on satisfactory explanation of the meaning of the terms. If a theory claims to furnish a safe bearing value, or if it operates with the coefficient of internal friction of clay, one may as well stop reading, unless the author explains in detail what he means by these terms.

The second requirement for an acceptable theory consists in the presence of adequate evidence for the assumptions. If these assumptions were obtained by a radical simplification of reality, which is the rule in connection with theories pertaining to soils, the evidence for the results must be presented. Whatever evidence is available can be classed into one of the following five categories:

(a) No evidence whatsoever;
(b) Evidence obtained by distorting the facts;
(c) Unbalanced evidence; that is, evidence obtained by eliminating all those facts which do not sustain the claim;
(d) Inadequate evidence, covering the entire range of present knowledge, yet insufficient to exclude the possibility of a subsequent discovery of contradictory facts; and
(e) Adequate evidence.

No honest business man and no self-respecting scientist can be expected to put forth a new scheme or a new theory as a 'working proposition' unless it is sustained by at least fairly adequate evidence. In any case, we expect him to inform us on the uncertainties involved. Therefore it is surprising to find upon closer scrutiny that many of the accepted rules of foundation engineering are based either on no evidence whatsoever, or on unbalanced evidence, and that the textbooks do not mention this serious failing. These rules seem to pass from one generation of textbooks into the next by a process of diffusion, whereby the scruples regarding the inadequacy of the evidence disappear ...

Outlook. The skeptical attitude towards our conceptions, and the readiness to modify them in accordance with increasing knowledge of the material, must be considered the second outstanding achievement of soil mechanics. By patient observation we have learned to discriminate between what we really know and what we merely believed. The amount of knowledge sustained by adequate evidence is appallingly modest, and the number of factors with a decisive influence on soil behavior is very much greater than was expected twenty-five years ago. The successful analysis of the reaction of the earth to changes produced by loading or by excavation was paid for by a heavy sacrifice of simplicity. Moreover, the severe restrictions on further progress along purely theoretical lines have become obvious. One of the most instructive examples of these limitations is to be found in the theory of arching in soils behind the timbering of cuts. The theory demonstrates that arching develops. It discloses the mechanics of arching, and reveals the limits between which the distribution of the lateral pressure of the earth may range. At the same time it leaves no doubt that the real distribution of the pressure depends on the method of constructing the timbering. Since we are not in a position to evaluate this influence on the basis of abstract reasoning, we are obliged to secure the required information by direct measurement of the pressures in full-sized cuts. We face a similar situation in almost every other field of soil mechanics. Our advanced knowledge of the mechanics and physics of soils makes it possible to grasp most of the essential factors which govern the stress and strain and the equilibrium of real earth. It has brought to us a realization of the extremely narrow limits of the validity of the older theories, and informs us of the existence of sources of danger which previously were hardly suspected. Nevertheless, in order to make the step from the qualitative appreciation of what is going to occur to a quantitative forecast requires accurate and systematic observations on full-sized structures.

Foremost among the sources of error requiring thorough investigation is the difference between the soil in its original state, and after it is delivered in the laboratory. In some cases the correction for the errors produced by the effect of sampling and handling can be made by computing the deformation of the subsoil for earlier stages of construction, and subsequently comparing the results with those of direct measurement. The first volume of the Proceedings contains a very instructive example for a successful operation of this kind.

Since we have achieved a reasonably clear conception of the possibilities and limitations of future research, the function of this Conference is simple. It consists essentially in establishing personal contacts between those who are interested in the subject from a theoretical or a practical point of view, and in stimulating exchange of experience. Though it originated not more than twenty-five years ago, soil mechanics is already old enough to have acquired the modesty which springs from experience. We know today that nothing worth while can be accomplished in this discipline without the intelligent and patient cooperation of the practicing engineer in the field. Some of the most valuable contributions to the Proceedings are a direct result of such co-operation. For this reason, we are very happy to welcome among the guests of the Conference a great number of outstanding executives and experienced construction engineers. Since these men owe their success and their professional standing to a keen discrimination between reality and fiction, I am sure they will appreciate our feelings against half-baked textbook wisdom, and assist us in getting down to tangible facts."

After the opening session, Karl von Terzaghi spent the evening with two acquaintances drinking champagne. On Saturday, June 20, busses carried members, their hosts and members of families to West Point. The route followed the famous Westchester County Parkways, passing the Kensico Dam and Croton Lake, crossing the Hudson River over the Bear Mountain Bridge to the United States Military Academy. The return trip was made via the George Washington Bridge, whose bridge structure was inspected. On the next day the participants of the conference started for an excursion to New Jersey and Manhattan. In the afternoon they left for Boston as the conference was going to continue at Harvard University from June 22 until June 27. During the conference Karl lived in a Harvard dormitory in a splendid, quiet park on the Oxford Street. The meetings lasted around six hours everyday; nearly all participants were present. Some of them are mentioned in his diary, at times in a very negative tone: "The Egyptian pair Hanna and Tschebotarioff brought the atmosphere of Egyptian plots to Boston. The director of the school has done nothing at all already for years. Raes from Belgium, University Ghent, insignificant theorist. Bakenhus, cultivated, gentleman, skillful negotiator. Buchanan, US Waterways, ... Bony Texas face, lean, happy blue eyes. W.P. Creager, restless, overabundant friendly businessman. Capt. Dean, Chief Engineer, New Orleans district, factual, earnest, officer and competent practical man. Harza, tall figure, great head, dishonesty stands in his face. Hathaway, US Eng., Kansas City 'You are all wrong!'. One track mind. Housel, as if one has kneaded thoroughly his head with the blond thin beard. Pathological and nasty. Zimmermann, impersonal, made no impression. Kimball, Professor, a puffed up zero. Krynine, like an offended mother's boy. Needs a quarter of an hour for the smallest remark. Middlebrooks, Fort Peck Dam – a suspicious number! Proctor, slimmer than 8 years ago, serious interest – for business reasons. Winterkorn, with small scar, chemical treatment of soil, Un. of Missouri, Colombia Mo., the German intriguer. And then, last but not least – representative of Austria – government surveyor Tillmann. 'Who is that monkey with the medal?' "

The sessions were strenuous but not stimulating for Karl von Terzaghi and discussions inconsolable and sterile. It seems that only the discussions with friends in the evening, after the sessions, were of interest to him or such sessions where he was involved with discussions and his own lectures. One of these was his contribution: *The shearing resistance of saturated soils and the angle between the planes of shear.* This lecture, published in the proceedings of the Harvard conference, was probably his best contribution to soil mechanics because in this paper he brought the concept of effective stresses (see Section 2.1) to the attention of the engineers: "All the measurement effects of a change of the stress, such as compression, distortion and a change of the shearing resistance are exclusively due to changes in the effective stresses,

..."; and it was the formulation of this concept – the main principle in soil mechanics – which, among other achievements, made him famous in engineering circles. In the course of the preparation of an anniversary book for Karl von Terzaghi, L. Bjerrum, the highly talented and intelligent director of the Norwegian Geotechnical Institute (N.G.I.) wrote in a letter of May, 1958, to A. W. Skempton, another disciple of von Terzaghi: "I agree with you that the understanding of the importance of effective stresses and pore pressures is Terzaghi's most important work and that this fact should be appreciated in his Anniversary Volume." However, obviously von Terzaghi still had confused ideas on the concept of effective stresses. In the aforementioned letter, L. Bjerrum reported to A. Casagrande: "Dear Arthur, I have just received your letter of May 26, (1958, the author) concerning Terzaghi's reports and the abstract of his paper on the shear strength of cohesive soil (for the anniversary volume, the author).

Concerning Terzaghi's abstract, he showed it to me when we met in Crete some weeks ago. It contains a good deal of confusing information. In the first place, Terzaghi has tried to explain the development leading up to Hvorslev's test results. In this connection he involves himself in a very confusing concept concerning pore pressure." And indeed Karl von Terzaghi's concept suffered from a considerable drawback; for it was developed for the mixture body and not for the partial solid. Thus, all interaction effects were not considered.

Karl von Terzaghi's speech at the end of the conference was enthusiastically received. In his closing address he drew some conclusions. "First of all, the Conference gave us a bird's-eye view of the activities in the field of our endeavors, which could not possibly have been otherwise obtained. Although I never failed to keep in close contact with new developments, I was amazed at the wealth of useful information which had escaped my attention, and I am sure that all the other members feel the same way about it. The second outstanding achievement of the Conference is that it has given a powerful impetus to further observation and research.

This impetus is essentially due to the enthusiastic co-operation on the part of the delegates of twenty different countries scattered all over the globe. This joint effort represents one of the most inspiring manifestations of internal co-operation I have ever experienced. Therefore I wish first of all to express to the delegates of the foreign countries my heartiest thanks and appreciation."

Moreover, he pointed out "Final meeting on Thursday 25. Unanimous decision to continue the Conference as an international organization under my chair."

The results of the Conference – opening and closing addresses, talks and discussions – are contained in the three volumes of the Proceedings of the International Conference on Soil Mechanics and Foundation Engineering (Casagrande, 1936). Karl von Terzaghi did not agree with all the contributions. He criticized the contents of the proceedings with sarcastic words in his diary entries of August 20 and 21. "Finished Conference work. Looking through proceedings: a madhouse, thousands of unripe brains released for new ground" and "Finished study of proceedings. As if one has released a madhouse. Weak brains, without overview, work like bees, without being able to distinguish between truth and error, theory and hypothesis."

The daily life of Karl and Ruth went on normally. On September 5, this was happily interrupted: Ruth gave birth to their son Eric. "A strong boy, born at 8:12, mother healthy and sleeping."

On September 10, Karl von Terzaghi left Boston in order to lecture at various universities and to visit construction sites in the western part of the United States. His schedule included talks at the headquarters of the US Reclamation Service in Denver, Colorado, at universities in Washington, Idaho and California, the California Institute of Technology in Pasadena, the headquarters of Californian Road Construction in Sacramento, and in the American Societies of Civil Engineering in San Francisco and Los Angeles. He was back in Boston on October 19, 1936, and on October 30 the von Terzaghis arrived with the steamer *Hamburg* in Cuxhaven. Via Berlin they travelled to Vienna and on Sunday, November 1, 1936, they were back in Vienna, this time however, with little Eric.

The month of November ran quietly, without any notable events. Karl and Ruth visited friends and theater performances, e.g. Shaw's *Jeanne d'Arc* with the famous Paula Wessely in the title role. On November 24, Karl von Terzaghi attended a talk in *The Österreichische Ingenieur- und Architekten-Vereines* in which Professor Tillmann, member of the Austrian delegation of the First International Conference on Soil Mechanics, reported about the soils conference in Cambridge. After the talks, Professor Fillunger stood up "and said that other people have another opinion about soil mechanics and that he will have the opportunity very soon, to express himself on this subject. I answered briefly, though sharply, under general laughter." There is no evidence for the authenticity of this incident. The journal *Zeitschrift des Östereichischen Ingenieur- und Architekten-Vereines* contains only the talk of Professor Tillmann but, however, no discussion contributions.

Apart from the above mentioned warning, Karl von Terzaghi could not foresee that he would soon be violently attacked by his colleague Fillunger, nor that this attack would paralyze his ability to work for half a year.

3.2 The Founding of Modern Theory of Porous Media by Paul Fillunger

Soil mechanics deals, in general, with the description of the motion and thermal behavior of sand and clay, completely or partly filled with groundwater. From the mechanical and thermomechanical point of view the saturated and unsaturated soils represent very complex heterogeneously formed bodies with solid and liquid/gas constituents. In the past liquid- and gas-filled porous solids had been treated as one-component bodies without an internal structure (see von Terzaghi, 1923). However, clay and sand bodies are structured. They can contain closed and interconnected pores, whose geometry is, in general, not known. Moreover, if the pores are filled, e.g., with water, there is interaction between the porous body and the water, as a result of the different velocities and material properties of both constituents. This makes the description of the mechanical and thermomechanical behavior difficult.

At first sight, it seems to be possible to use two different strategies to solve the complex problem of saturated and unsaturated porous bodies under loads or heat. For this purpose, it would be necessary to separate the constituents. Then, the axioms of mechanics and thermodynamics would be applied to the separated constituents in consideration of all interaction agencies. Such a procedure may, in special cases, be successful. However, for the development of a general theory of saturated porous solids, this procedure is completely unsuitable because of the complex internal geometric structure of the pores of the solid skeleton. Moreover, for many problems, this procedure is not at all necessary, since these problems require a macroscopic description of physical phenomena rather than a description on the microscopic scale; e.g., for the description of seepage, the planning engineer is not interested in the real velocity of the water in the individual pores of the porous solid. Instead, for his design he needs the seepage velocity, i.e. the average velocity of the water flow in all pores. Thus, it depends on the problem whether investigations in microscopic or macroscopic ranges are necessary, whereby the macroscopic range is as the range where all quantities in the microscopic range are settled as statistical average values.

The above considerations show that a second strategy is needed, which can be found in the macroscopic description of the problem. Thus, it is convenient to describe saturated and unsaturated porous solids by means of a substitute model consisting of heterogeneously composed continua with internal interactions. This procedure involves the concept of volume fractions to distribute the mass of the solid skeleton and of the liquid over the total control space, which is shaped by the porous solid. The distribution takes place with the help of volume porosity numbers which fix the ratio of the volumes of the individual constituents in relation to the volume of the control space, which is shaped by the porous solid. This procedure assumes indeed that the pores are statistically distributed over the control space. In this case, the equality of volume and surface porosity is given as a statistical necessity. The concept of volume fractions has the effect that *smeared* substitute continua with reduced densities for the solid and liquid phases arise which fill the control space simultaneously and which can be treated with the methods of continuum mechanics. It is obvious that the same properties which characterize the mixture theory appear, for, in mixture theories, it is also assumed that the individual parts of the mixture cover the total control space.

The purely mechanical mixture theory was developed by Stefan (1871) and has come to well-founded conclusions. Thus, with the continuum theory of mixtures, containing various constituents with their own degrees of freedom, an ensured basis is available to treat the mechanical behavior of saturated and unsaturated porous solids.

This idea was recognized by Fillunger (1936) when he tried to improve upon von Terzaghi's attempt to treat saturated clay under load, in particular, von Terzaghi's and Fröhlich's (1936) attempt to describe the consolidation of soil. Fillunger was the first scientist to put the whole procedure on a scientific basis [47].

Fillunger (1936) studied the consolidation problem for six months. Then he published his results in the second section of the brochure, *Erdbaumechanik?*. He proceeded from a two-phase-system, which he described with ensured mechanical axioms and principles:

"One easily realizes that for the theoretical investigations of the settlement due to groundwater flows, the Eulerian fundamental equations of hydrodynamics must be applied. It would not cause any difficulties to set them up immediately under possible general assumptions; however, it is recommended to treat only the simplest case at first, for which all processes depend only on the time t and on one single coordinate z. It is then easier for the reader to examine the derivation of the equations concerning their correctness."

Then, Fillunger started immediately to develop the fundamental equations of porous media theory. He stated that it was a matter of two coupled flows concerning the porewater flow and the settlement: "The porewater (body 1) flows upwards, and the porous soil (body 2) flows downwards with the settlement rate. If we, like the other authors (von Terzaghi and Fröhlich, the author), disregard the effect of own weight, then, the external force for each body consists only of the resistance to this flow put up by the other body, and the coupling of the two motions is based on this. It is further recommended that this external force no longer be related to the mass unit but rather to the unit of volume, and that it be imagined that the porewater constantly, but with varying density, fills the total space as well as the soil. It is then as if two flows exist in the same space, two flows which can influence each other only through the flow resistance, but not according to the law of space-displacement."

---

[47] The following paragraphs are for experts. Readers who are not interested in the development of the porous media theory can proceed immediately to Section 4.

$$
(7) \quad \ldots\ldots \left\{ \begin{aligned}
&\frac{\partial v_1}{\partial t} + v_1 \frac{\partial v_1}{\partial z} = \frac{1}{\varrho_1}\left(-Z - \frac{\partial p_1}{\partial z}\right),\\
&\frac{\partial v_2}{\partial t} + v_2 \frac{\partial v_2}{\partial z} = \frac{1}{\varrho_2}\left(Z - \frac{\partial p_2}{\partial z}\right),\\
&\frac{\partial \varrho_1}{\partial t} + \frac{\partial (\varrho_1 v_1)}{\partial z} = 0,\\
&\frac{\partial \varrho_2}{\partial t} + \frac{\partial (\varrho_2 v_2)}{\partial z} = 0.
\end{aligned} \right.
$$

Fillunger's basic equations

After this statement, Fillunger (1936) wrote down the balance equations of momentum and mass for both bodies for the one-dimensional motion of a binary model, excluding any mass exchange.

In these equations $v_{..}$ denotes the velocity parallel to the z-axis, $\rho_{..}$ the density, $p_{..}$ the pressure of the single constituents, and Z the interaction force between the porewater and the soil.

In the following, Fillunger (1936) admitted only a hydrostatic pressure, which he identified with the real liquid pressure, and related this pressure, via Deless' law and the volume fraction n, to the partial solid and liquid phases. Moreover, he considered only incompressible constituents and reformulated his basic equations.

"In each cross-section z of the double-flow, there exists a pressure p which is distributed amongst the two bodies or materials as follows: If n is the pore space per unit of volume, then it follows from the Delessian law (where on each cut surface, in a uniform mixture, the surface ratios of each partial constituent must be equal to their volume ratios) that the partial pressure must be decomposed into p_1 (porewater)

$$p_1 = np\ ,$$

and into the partial pressure p_2 (clay phase)

$$p_2 = (1-n)p\ .$$

In fact, $p_1 + p_2 = p$. If we denote the constant density of the incompressible porewater with ρ_1', its specific weight with γ_1, as well as the constant density of the solid particles in the clay (which are also considered as being incompressible) and its specific weight with ρ_2' and γ_2, then (acceleration due to gravity $= g$)

$$\rho_1' = \frac{\gamma_1}{g} \quad \text{and} \quad \rho_2' = \frac{\gamma_2}{g}\ .$$

In the distributed state, related to the total space, we then have

$$\rho_1 = n\rho_1' = n\frac{\gamma_1}{g} \quad \text{and} \quad \rho_2 = (1-n)\rho_2' = (1-n)\frac{\gamma_2}{g} \,.$$

If one introduces the partial pressures as well as the densities according to these relations into Eqn. (7), (see upper picture, the author) taking into consideration that both p and n are functions of t and z, then after dividing by the constant factor $\frac{\gamma_1}{g}$ or $\frac{\gamma_2}{g}$, respectively, one obtains:

$$\frac{\partial v_1}{\partial t} + v_1\frac{\partial v_1}{\partial z} = \frac{g}{n\gamma_1}\Big(-Z - \frac{\partial(np)}{\partial z}\Big) \,,$$

$$\frac{\partial v_2}{\partial t} + v_2\frac{\partial v_2}{\partial z} = \frac{g}{(1-n)\gamma_2}\Big(Z - \frac{\partial(1-n)p}{\partial z}\Big) \,,$$

$$\frac{\partial n}{\partial t} + \frac{\partial(nv_1)}{\partial z} = 0 \,,$$

$$-\frac{\partial n}{\partial t} + \frac{\partial(1-n)v_2}{\partial z} = 0 \,.\text{”}$$

These are the material-independent fundamental equations – within the framework of a purely mechanical theory for a one-dimensional motion – of a binary model consisting of an incompressible liquid and an incompressible solid skeleton. Fillunger (1936) consequently used the balance equations of momentum and mass excluding any mass exchange, separately for both constituents, whereby he included, in the balance of momentum, interaction forces Z as volume forces. Subsequently, he determined the interaction force, with the help of Darcy's law, where k is the coefficient of permeability.

$$Z = \frac{v_1 n}{k} \,.$$

The drawback of this derivation was entirely known to him, for he later indicated that the above equation had been derived with the assumption of a stationary and a homogeneous flow and a rigid solid skeleton with the velocity v_2 equal to zero and the porosity n constant: “It is obvious that in the treatment of the porewater flow which is coupled with the settlement, v_1 must be replaced by the relative velocity $(v_1 - v_2)$. However, one has also to consider that $n = const.$ in the filtration test.”

Even if one cannot agree with all of Fillunger's further arguments (1936), four remarks in view of “desirable generalizations” are of interest: “One easily realizes which generalizations are desirable and even partly necessary in order to form a theoretical tool applicable in the field of civil engineering. They may be briefly listed here:

1. Introduction of x, y, z as coordinates instead of z alone, since buildings immediately load only a small region of the earth surface. Then, it may become unavoidable to consider the internal friction of viscous liquids and to incorporate additional terms in the motion-resistances which must contain the velocity gradients and a special friction constant.

2. Introduction of further permeability coefficients, e.g., besides k' for vertical flow, k'' for horizontal flow because the arrangement in layers can suppose different permeabilities in these two directions.

3. Introduction of water's own weight and that of solid clay particles as additional terms to $-Z$ and $+Z$, ... Then, also the so-called 'sedimentation' could be incorporated, i.e., the natural deposition and hardening of mud.

4. If the clay particles have a compression resistance similar to the elastic body, then in the second (of the reformulated equations, the author), and only there, an additional term to Z of the form $\frac{\partial f(n,k''')}{\partial z}$ should be added. This means that $p = f(n, k''')$ is an equation of state, p is a pressure (kg/cm^2) and k''' is a new material-dependent value. The form of the function $f(n, k''')$ and the constant k''' would have to be determined from tests, which we could confidently call ödometer tests."

The most important generalization in Fillunger's list is the fourth point, wherein he formulated masterfully and explicitly in a completely correct way the concept of effective stresses for the partial porous solid, and not for the mixture body as von Terzaghi had done.

In a footnote, Fillunger (1936) pointed out that von Terzaghi and Fröhlich (1936) had obviously not needed any extension of their relations in order to calculate the sedimentation.

Contrary to von Terzaghi's (1923) procedure, Fillunger's (1936) approach corresponds to the modern concept of the treatment of porous media, namely to the mixture theory restricted by the volume fraction concept. It represents an extraordinary scientific achievement of high originality when one considers that, at that time, the modern mixture theory had not yet been developed, much less a porous media theory.

4. The Scandal

4.1 The Run of Events

In 1933 and 1934, the previously-mentioned dispute with von Terzaghi over the uplift and capillary problems had arisen and, on February 19, 1935, Fillunger heavily criticized a report by Lehr (1934) in a speech to *The Österreichische Ingenieur- and Architekten- Verein.* This speech was filled with many polemical statements and caused such a sensation that a discussion was arranged to take place on April 5, 1935. The discussion evening ended with a final remark by Fillunger in which he concluded that his 25 years of teaching about the strength of materials qualified him to criticize Lehr's (1934) contribution. Moreover, in the wake of the controversy, the rector of the Technische Hochschule was obliged to intervene. The *Österreichische Ingenieur- and Architekten- Verein* refused the publication of Fillunger's talk in its own journal, as was usual, with the excuse of lack of space. Therefore, Fillunger edited the talk by himself (Fillunger, 1935). In seeking to clarify the truth, he continued to attack Lehr's report (1934). In December, 1935, Lehr's report (1934) was advertised in the journal, *Zeitschrift für Angewandte Mathematik und Mechanik*, which was, at that time, edited by Professor Trefftz. In two letters of January 7 and 21, 1936, to the editor, Fillunger intervened against the advertisement. In the last letter, he mentioned that not only the "realistic strength of materials" of Lehr (1934) had to be criticized extensively, but also "a certain soil mechanics." It then also seems that in this letter Fillunger announced for the first time his objective to attack von Terzaghi's work.

The two publications of von Linsemann (see Fillunger, 1935) and Foerster (see Fillunger, 1936) finally convinced Fillunger to act as a rigorous reviewer in new fields of technical science.

"Whoever walks along with insolent steps, dresses up with strange feathers without thanking those from whom he learned or borrowed, who digs his claws into everything, who accepts nothing other than himself, who belittles every opponent or predecessor, who combines his predominant arrogance with incompetence, falls into the hands of only a courageous censor who carries a strong scourge and can protect the peaceful doves against the croaking vulture. It is sufficient that in all the *superfluous knowledge*, the stale vanity

Theorie

der

Setzung von Tonschichten

Eine Einführung in die analytische Tonmechanik

Von

Dr. Ing. K. v. Terzaghi und Dr. Ing. O. K. Fröhlich

o. ö. Professor an der Technischen Hochschule in Wien

Beratender Ingenieur in 's-Gravenhage

Mit 100 Abbildungen im Text

Leipzig und Wien

Franz Deuticke

1936

The book *Theory of the settlement of clay layers*
by Karl von Terzaghi and Otto K. Fröhlich

of modern education governs such a wide region and one is obliged to throw some stones into the stagnating, swampy water.

If then, however, the glistening emptiness is enthroned in order to give the literature rules and the morals commands, a counterthrust is necessary which is strong enough and loud enough to be noticed by the better people in human society, and also strong and strict words are necessary.

The reviewer has not only to fight an open and honest battle with the author, but rather also another one with invisible powers which intend to mislead opinion."

"Who does not know the virtuosos and swindlers of the word against whom the human being, who deliberately once says an untruth, appears primitive and harmless? The so-called 'aesthetic', but also the scientific literature of Germany, was not exactly poor in those; and is also still not so today. With extremely little of their own substance of intellect, spirit, mind, emotion, belief, immediate knowledge and experience, these liars initiate an amazingly wealthy vocabulary. The language has apparently lost its roots and performs in a ghostlike manner an independent life; it is the applied technique of pure talk, is recollection, an association and reproduction function. A horrible, idle motion of the words sets in: rich, glittering, dazzling, pompous, astonishing – and hollow."

In the background of the aforementioned polemical discussions with the consequences drawn by Fillunger, an occurrence which is unique in European scientific history must be evaluated. Obviously, Fillunger was so deeply hurt by Stern's, Hoffman's, and, in particular, von Terzaghi's attacks, that he himself became the aggressor. As already indicated in a letter to Professor Trefftz, he intended to criticize extensively "a certain soil mechanics". The opportunity to do this came shortly afterwards. In the spring of 1936, Fillunger made an effort to obtain permission to publish a book on material strength with the publishing house Deuticke – an effort which subsequently failed. On this occasion, the book *Theorie der Setzung von Tonschichten* (Theory of settlement of clay layers) by von Terzaghi and Fröhlich (1936) was shown to him for the first time. Fillunger bought the book and studied it extensively during the summer semester. He came, finally, to the conclusion that he must refuse the book on scientific grounds. In the time to follow, he developed the fundamental equations of the purely mechanical porous media theory; he had succeeded in this by the end of September, 1936. In October, 1936, he began to write down his criticism of von Terzaghi's work, especially of the book by von Terzaghi and Fröhlich (1936), and he represented his own view of the description of settlement problems.

He decided to publish his criticism and his investigations in a pamphlet which he edited himself, owing to the bad experiences he had with the publication of his polemics against von Terzaghi. In the course of these polemics, some of Fillunger's contributions had been refused previously by the editors, as had already been mentioned. Thus, in editing the brochure by himself, he saw the only possibility to publicize his ideas.

The public had first learned of his objections through rumors during the summer of 1936. Then, at the end of the summer semester, Fillunger visited his colleague Professor Lechner, who had formerly been a co-worker of the professors Jaumann and Hamel at the Deutsche Technische Hochschule in Brno, and declared that he had discovered severe mistakes in the book by von Terzaghi and Fröhlich. Fillunger also repeated his objections to the "Dozent", Dr. Magyar (fluid mechanics). Finally, in October, 1936, Fillunger announced to Professor Lechner, during a bus ride, that he would hold a talk on soil mechanics in *The Österreichischen Ingenieur- und Architekten-Verein, Wien.* On this occasion, Fillunger stated that one had to maintain order at the Technische Hochschule of Vienna or one would be driven directly to Bolshevism. Moreover, it would also be necessary to attack other professors. However, since at this time he had been too involved in the affair with von Terzaghi, he was forced to hold back on this subject.

Fillunger started his last attempt to represent his view on soil mechanics in a talk in *The Österreichischen Ingenieur- and Architekten- Verein* on October 23, 1936. The talk, entitled *Erdbaumechanik ?*, was at once declared unwelcome by the general manager, Ing. Willfort. However, it was decided that the governing board should be consulted. A spontaneous counterproposal

O. K. Fröhlich:
Life-time friend of Karl von Terzaghi
and his successor in Vienna

from the general manager was accepted by Fillunger. Ing. Willfort proposed that a talk should be held in which Fillunger would speak for the first 45 minutes, and then von Terzaghi would have his turn. In addition Fillunger was to hand his explanation over in written form to von Terzaghi. Fillunger agreed to this proposal. However, in a later discussion, Ing. Willfort did not uphold his own offer. Finally, the governing board declined to hold the talks on November 19, 1936. Fillunger was informed orally and then in writing (November 25, 1936).

Thereupon Fillunger decided to distribute his pamphlet *Erdbaumechanik ?*, which had been printed in the meantime. The distribution took place on December 2, 1936. The brochure was sent to more than 100 different people and institutions all over the world, though, in particular, to those in Austria and Germany. He explained his procedure to the president of *The Österreichischen Ingenieur- und Architekten- Verein*: "It is here a matter of a vital question concerning technical science, as I have already stated in my talk of February 15, 1935. It would not make any sense to give real technical-scientific teaching if the areas which we develop further are contaminated by such theories. Professor means, however, 'confessor', and for this reason one must reach for

ERDBAUMECHANIK?

VON

O. Ö. PROF. DR. PAUL FILLUNGER
WIEN

IM SELBSTVERLAGE DES VERFASSERS
DRUCK UND VERTRIEB DURCH DIE
BUCHDRUCKEREI FRIEDRICH JASPER, WIEN, III., THONGASSE 12
(1936)

The polemical brochure
Earthwork mechanics? by P. Fillunger

decisive and prompt defense as soon as such great errors (to put it mildly) have been recognized ...

Allow me, Mr. President, to make a remark in conclusion. It will surely be attempted to accuse me of being perfidious because I have urgently written my wish to give a talk, without simultaneously informing either you or general manager Mr. Willfort, of my objectives and at which stage of performance they are already positioned. I did not believe that I would be obliged to do this. The severity of the battle, which I am forced to fight with very unequal weapons, doubtlessly justifies the fact that I might not give away the little advantage of being able to face a not so well-prepared opponent. Also, I have done everything in my power to not leave him in the dark about my objective to attack him."

On Thursday, December 3, 1936, von Terzaghi was surprised by the news that Fillunger had started an attack against him and against soil mechanics as a science. Von Terzaghi's friends and associates were in a great panic. In the evening of the same day, he received a copy of Fillunger's brochure.

Fillunger's (1936) pamphlet, *Erdbaumechanik ?*, contained six main parts. In his brief introduction, Fillunger stated that the consolidation theory of von Terzaghi and Fröhlich (1936) was in no way satisfactory, and had to be declined. Due to the eminent importance of the theory of settlement for civil engineering, however, Fillunger continued, a scientific-critical treatment of this subject was justified by another author. Then, Fillunger started his comments, which were a mix between technical criticism and personal defamation.

In the first section of his pamphlet, he showed briefly the derivation of von Terzaghi's partial differential equation of the consolidation problem from 1923. He accompanied nearly each step of the derivation with comments, and remarked finally that the acceleration was missing in von Terzaghi's investigations. Moreover, he criticized von Terzaghi's *thermodynamisches Gleichnis* (thermodynamic parable).

In the second section, he developed his own ideas regarding the consolidation problem, in which he founded the purely mechanical theory of porous media (see the explainations in Section 3.2).

Then, in the third part of his pamphlet, Fillunger vehemently criticized the integration of von Terzaghi's differential equation performed by von Terzaghi and Fröhlich (1936). In particular, he criticized single calculation steps and accompanied them with sarcastic remarks.

Soil-mechanical tests in the laboratory were the target of Fillunger's criticism in the fourth passage of his self-edited brochure *Erdbaumechanik?*. After some general critical remarks about experimental research, he tried to prove that von Terzaghi and Fröhlich (1936) had used a particular form of a total differential which was denoted by Fillunger as the "Terzaghi-Differential." In this specific point, Fillunger was in error, as will be seen later. He also criticized single experiments, in particular the ödometer test with disturbed soil samples. Finally, he attacked the tests on the shear strength and on the internal friction of soils.

In the fifth section, concerning applications in constructional engineering, Fillunger's criticism culminated with the statement: "Applications in constructional engineering! There are not any if one attempts by this statement to conclude the unpleasant topic once and for all. One could only simply add: there will never be any applications, at least not as long as the construction of houses, roads and bridges takes place in the hitherto usual way on the solid surface of the continents."

The last section, finally, contained some very general remarks and statements, and it was particularly in this section that Fillunger justified his heavy criticism.

As already-mentioned, the technical comments were accompanied to a great extent by personal defamation. The main attacks are summarized in the following:

"It is difficult to bear the idea that an examination candidate, who was probably better able to manage this mathematical-geometrical problem than the authors, could therefore have suffered a delay in completion of his studies or even their premature failure! ... "

"The throwing around of scientific terms like this must of course leave an extraordinarily auspicious impression on people of the same fictitious scientific education of the most superficial kind."

"Jugglers must, if they want success, exhaust and divert the attention of their audience. With such scientific juggling, as this Fröhlich- [48] Terzaghi-mathematics is, readers and students shy away from what they could show by unfounded criticism or already by pure confession: This is what I do not understand! A severe lack of knowledge and lack of keen intellect is added in, which is favorable to the success of the jugglers. However, one cannot excuse the leading scientific authorities, the professors, who were technically not complete strangers to these things, who observed these from the very beginning without protesting even promoting or in some other way participating ..."

"One should think little of the frankness with which the 'obvious faults' are confessed, the 'most important sources for mistakes' are listed in the bombastic preface, whereby the most important lack of knowledge and the ability of the authors are, of course, not mentioned and that the entirety only means a 'first approach'. This is already well-known."

The statements (not all are cited here) that Fillunger printed in his pamphlet represented in totality the scandalous accusation that von Terzaghi had invented a dummy science and had betrayed all professional circles through a clever swindle, all for the purpose of self-enrichment. This was implied to have begun in his teaching program at the universities and subsequently ending with the *International Congress for Soil Mechanics* at Harvard in 1936.

Immediately after the appearance of the pamphlet *Erdbaumechanik ?*, von Terzaghi embarked upon great activities. Already on December 4, 1936, he had had discussions with his colleagues and friends Professor Orley and Schaffernak, specialists for road, railway, and tunnel construction, as well as hydraulics. Moreover, he was accepted by the Rector of the Technische Hochschule of Vienna, Prof. Dr. techn. Friedrich Böck, and he received several telephone calls concerning Fillunger's attack, including one from the general manager of the *Österreichische Ingenieur- und Architekten-Verein*, Willfort. In the afternoon of December 5, 1936, Dr. W. Herold visited von Terzaghi. Dr. Herold had written a book concerning some strength problems which had been published by Julius Springer in 1934. This book was strongly and sarcastically criticized by Fillunger in his talk on *Alte und neue Probleme der Festigkeitslehre*, in the *Österreichische Ingenieur- und Architekten-Verein.* Von Terzaghi and W. Herold arranged that Herold should write a memorandum against Fillunger in order to weaken Fillunger's position. This was written the next day and then typed on December 12, 1936. In this memorandum, Herold inexorably criticized Fillunger's scientific work. In the time between December 5 and December 11, von Terzaghi studied Fillunger's pamphlet and worked, together with Fröhlich, on a rebuttal. The publisher, Deuticke, acquainted with von Terzaghi for a long time, promised to publish the reply in his publishing house.

[48] Fillunger used O.K. Fröhlich's name as an adjective, punning on the German word for *happy*, the author.

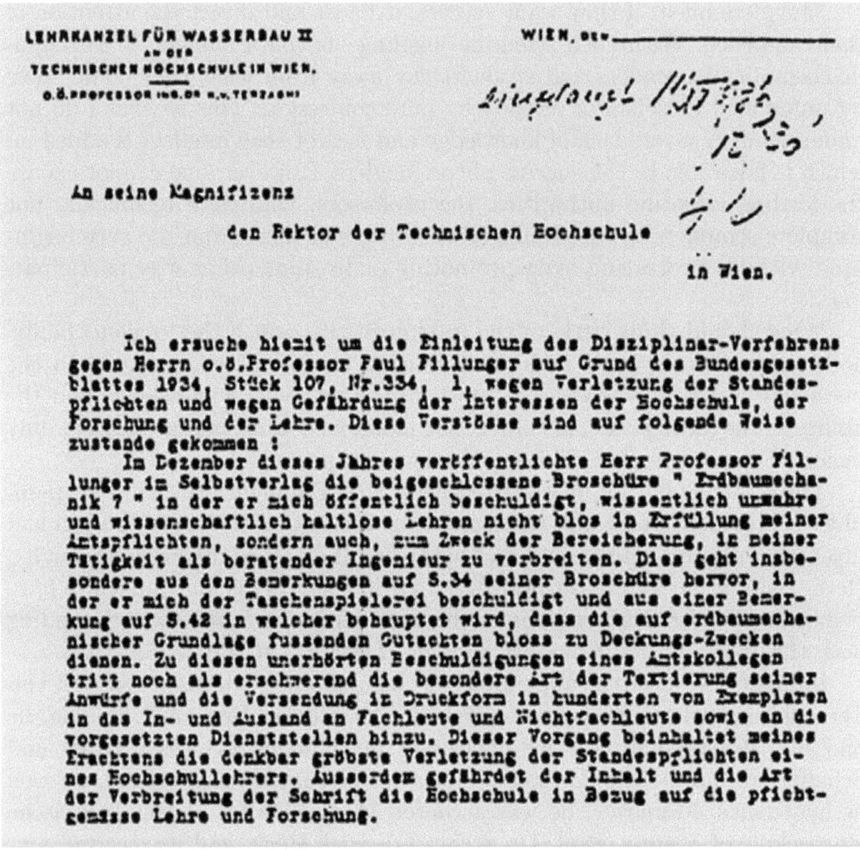

LEHRKANZEL FÜR WASSERBAU II
AN DER
TECHNISCHEN HOCHSCHULE IN WIEN.
O.Ö.PROFESSOR ING.DR. K.v.TERZAGHI

WIEN, DEN

An seine Magnifizenz

den Rektor der Technischen Hochschule

in Wien.

Ich ersuche hiemit um die Einleitung des Disziplinar-Verfahrens gegen Herrn o.ö.Professor Paul Fillunger auf Grund des Bundesgesetzblattes 1934, Stück 107, Nr.334, 1, wegen Verletzung der Standespflichten und wegen Gefährdung der Interessen der Hochschule, der Forschung und der Lehre. Diese Verstösse sind auf folgende Weise zustande gekommen :

Im Dezember dieses Jahres veröffentlichte Herr Professor Fillunger im Selbstverlag die beigeschlossene Broschüre " Erdbaumechanik ? " in der er mich öffentlich beschuldigt, wissentlich unwahre und wissenschaftlich haltlose Lehren nicht blos in Erfüllung meiner Amtspflichten, sondern auch, zum Zweck der Bereicherung, in meiner Tätigkeit als beratender Ingenieur zu verbreiten. Dies geht insbesondere aus den Bemerkungen auf S.34 seiner Broschüre hervor, in der er mich der Taschenspielerei beschuldigt und aus einer Bemerkung auf S.42 in welcher behauptet wird, dass die auf erdbaumechanischer Grundlage fussenden Gutachten bloss zu Deckungs-Zwecken dienen. Zu diesen unerhörten Beschuldigungen eines Amtskollegen tritt noch als erschwerend die besondere Art der Textierung seiner Anwürfe und die Versendung in Druckform in hunderten von Exemplaren in das In- und Ausland an Fachleute und Nichtfachleute sowie an die vorgesetzten Dienststellen hinzu. Dieser Vorgang beinhaltet meines Erachtens die denkbar gröbste Verletzung der Standespflichten eines Hochschullehrers. Ausserdem gefährdet der Inhalt und die Art der Verbreitung der Schrift die Hochschule in Bezug auf die pflichtgemässe Lehre und Forschung.

Karl von Terzaghi's request to open a disciplinary action

Moreover, Karl von Terzaghi considered bringing a slander suit against Fillunger. However, he received the urgent advice to renounce such an action from renowned lawyers. The lawyer Dr. Herrdegen remarked in a letter concerning von Terzaghi's request: "Now I have still to touch on a question which can be decisive in a trial. In reading the brochure the supposition presses whether one is dealing, in this case, with an author who is known as a querulous person ...

Because the court can convict an accused person only then, when he possesses the awareness of his complete responsibility, it cannot be ruled out that a psychological check-up would be in order and that the trial would be lost if there is a positive result."

During a meeting of the board of the Professorskollegium on December 8, 1936, Prof. Dr. Hartmann brought up *Earthwork Mechanics?*. The discussion led to the resolution bringing the disciplinary prosecutor into play.

On December 10, 1936, von Terzaghi formulated an urgent request to open a disciplinary action against Fillunger on the basis of a paragraph in the *Bundesgesetzblatt of 1934*, because of an alleged violation of Fillunger's class-duties and his jeopardizing of the interests of the Technische Hochschule. He included his request in a two-page letter to the Rector of the Technische Hochschule of Vienna. However, the Rector had already written, on December 9, to Professor Dr. Merkl, the *Disziplinaranwalt*, of the Technische Hochschule of Vienna, asking him whether an offense against regulations had been committed concerning the content of some statements in the brochure *Erdbaumechanik ?*. Prof. Dr. Merkl answered the next day at 11 p.m., declaring that Professor Fillunger's comments in his brochure exceeded the approved special scientific criticism allowed under constitutional law and gave reasons for the suspicion of an offense against regulations. Prof. Merkl left it to the Rector's discretion whether or not to lay a disciplinary action against Professor Fillunger at the *Disziplinarkammer* (Disciplinary Court). This was done by the Rector in a letter to the president of the *Disziplinarkammer*, Prof. Dr. List, on December 12, 1936. In a strictly confidential letter to the president of the *Disziplinarkammer*, at the Technische Hochschule of Vienna, Magnifizenz Prorector Professor Ing. Franz List, the Rector opened the disciplinary action, sighting the suspicion of an offense against regulations, in the form of an accusation against a colleague, of profit-seeking utilization of deliberately falsified theories, further sighting the deliberate toleration and promotion of this behavior on the part of other professors on the basis of the enclosed pamphlet *Erdbaumechanik ?* by Professor Dr. Paul Fillunger.

Furthermore, the Rector remarked that the disciplinary lawyer would, in the case of the announced disciplinary information by the Rector, seek for the opening of a disciplinary investigation against the o. Professor Paul Fillunger.

In a letter of December 15, 1936, Magnifizenz Prorector Prof. Ing. Franz List called for the *Disziplinarkammer* to assemble at 6 p.m. on December 21, 1936. Members of the "*Disziplinarkammer* für Bundeslehrer an der Technischen Hochschule in Wien" were at that time the following professors of the Technische Hochschule and the Universität of Vienna:

President:	Magn. Prorector Prof. Ing. Franz List
Members of the Senate:	o. Prof. Dr. Heinrich Mache
	o. Prof. Dr. Heinrich Pawek
	o. Prof. Dr. Friedrich Schaffernak
	o. Univ. Prof. Dr. Josef Hupka
Disciplinary lawyer:	o. Univ. Prof. Dr. Adolf Merkl
Substitutes:	o. Prof. Dr. Erwin Kruppa
	o. Prof. Dr. Engelbert Wist
	o. Prof. Dr. Karl Holey
	o. Univ. Prof. Dr. Ferdinand Degenfeld-Schonburg

In the meantime, the first protest letters arrived at the *Rectorat* of the Technische Hochschule of Vienna. In a letter of December 12, 1936, the president of the *Österreichische Ingenieur- und Architekten-Verein*, Ing. Brabbée, expressed his urgent request to the collegium of professors to do everything possible in order to clean up the whole affair. This was in the interest of the Technische Hochschule, as well as in the interest of the entire engineering industry.

The next protest letter was from Dr.-Ing. Otto Karl Fröhlich, the co-author of the book *Theorie der Setzung von Tonschichten*, which had been strongly attacked by Fillunger (1936). Fröhlich mentioned in his letter that in the mathematical part of the pamphlet, Fillunger had defamed him in a severe way, and that this had caused him substantial economic losses in his consulting work. Moreover, he stated that he was able to refute Fillunger in the application of the higher analysis of physical problems, and he requested the establishment of a committee of experts.

The Rector of the Technische Hochschule also received protest letters from abroad, namely from Mr. Cummings of the Raymond Concrete Pile Company (USA), from Mr. Proctor of the American Society of Civil Engineers (Soil Mechanics and Foundations Division), and from Mr. A. Casagrande, professor at Harvard University, Cambridge, Massachusetts. The most severe letter was written by A. Casagrande. He threatened the Technische Hochschule of Vienna that the International Conference on Soil Mechanics and the American Society of Civil Engineers were ready to intervene in the affair if the Technische Hochschule did not take action against Fillunger's defamations. Finally, A. Casagrande pointed out that von Terzaghi had planned to organize the next Second International Conference of Soil Mechanics in Austria. Undoubtedly, the decision on this proposal would depend strongly on the reactions of the Technische Hochschule and the Austrian engineers in the affair.

The activities of both von Terzaghi and Fröhlich to overcome Fillunger's brutal attack continued. After von Terzaghi had cancelled some lectures for the students and had handed others over to his co-worker Steinbrenner, he delivered a small address to his students at 9.45 a.m. on December 15, 1936. He began by explaining the conflict with Fillunger and then went on to make some general statements: "Successful teaching is based upon the students' respect for their teachers and upon the confidence of the students in the expertise of their professors. The aggressor has undertaken a large-scale attempt to sap your respect in my person and to shake your confidence in my expertise." Von Terzaghi announced that he would comment on the contents of the attack at a later date. He stated that until then he would have to turn his lectures over to his co-workers because he needed every minute of his time to defend himself against the attacks.

Indeed, von Terzaghi and Fröhlich worked without rest on their reply to Fillunger's brochure. They were also in contact with Dr. Leo Rendulic,

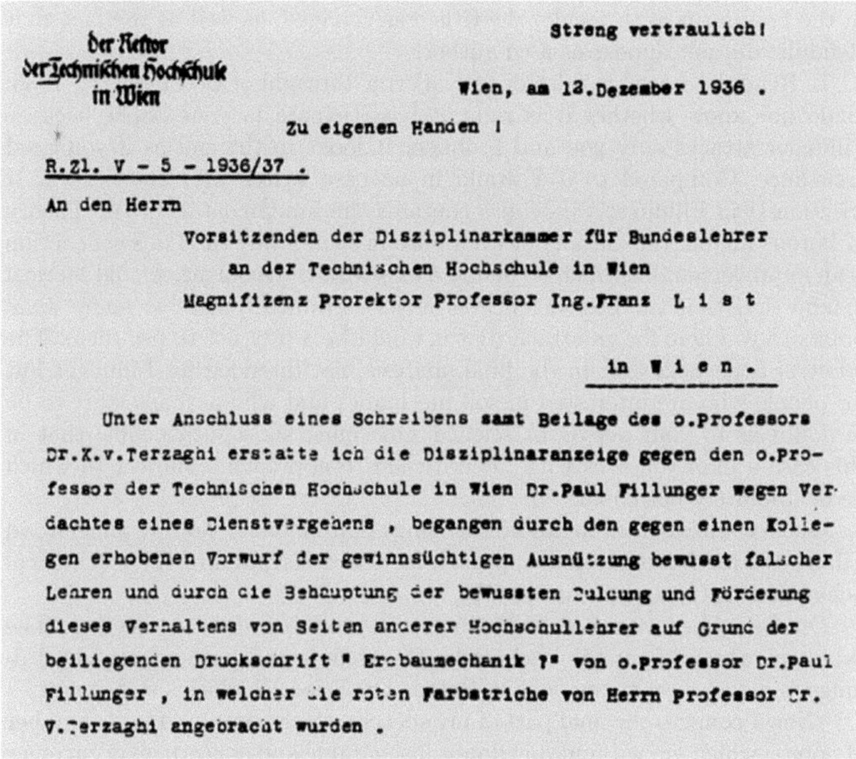

Der Rektor
der Technischen Hochschule
in Wien

Streng vertraulich!

Wien, am 12.Dezember 1936 .

Zu eigenen Handen !

R.Zl. V - 5 - 1936/37 .

An den Herrn
Vorsitzenden der Disziplinarkammer für Bundeslehrer
an der Technischen Hochschule in Wien
Magnifizenz Prorektor Professor Ing.Franz L i s t

in W i e n .

Unter Anschluss eines Schreibens samt Beilage des o.Professors Dr.K.v.Terzaghi erstatte ich die Disziplinaranzeige gegen den o.Professor der Technischen Hochschule in Wien Dr.Paul Fillunger wegen Verdachtes eines Dienstvergehens , begangen durch den gegen einen Kollegen erhobenen Vorwurf der gewinnsüchtigen Ausnützung bewusst falscher Lehren und durch die Behauptung der bewussten Duldung und Förderung dieses Verhaltens von Seiten anderer Hochschullehrer auf Grund der beiliegenden Druckschrift " Erdbaumechanik ?" von o.Professor Dr.Paul Fillunger , in welcher die roten Farbstriche von Herrn Professor Dr. v.Terzaghi angebracht wurden .

Disciplinary action of the Rector

the respected former collaborator of von Terzaghi, who had in the meantime moved to Berlin as already indicated. Fröhlich wrote to Rendulic that he and von Terzaghi wished to incorporate him into fighting the attack. He further remarked that the publisher Deuticke had offered to publish the report. The title of the brochure came to von Terzaghi's mind in a flash:

ERDBAUMECHANIK ?	SOIL MECHANICS ?
und	and
ERDBAUMECHANIK !	SOIL MECHANICS !
Eine	A
Entgegnung	Reply
von	by
K. V. TERZAGHI	K. V. TERZAGHI
Unter Mitarbeit von	In Cooperation with
O. K. Fröhlich	O. K. Fröhlich
und	and
L. Rendulic	L. Rendulic

In the final form of the reply, the title was changed, as well as the fact that Rendulic did not appear as a co-author.

L. Rendulic stated in a letter to Karl von Terzaghi of December 22, 1936: "I do not know whether it is right if I participate in your paper because Fillunger attacks only you and Fröhlich. It looks in the end as if you need assistance. Compared to it I would in no case switch over the Section II (Section II in Fillunger's brochure contains the fundamentals of the Theory of Porous Media, the author) or even pose in such a way that this concoction is an improvement for which no demand exists for the moment. An old tactical maxim says that the best defence is offense. Fillunger offers so many weak points to you here for an attack that it would be a pity not to use them. The rebuttal is nevertheless, in the final analysis, not intended for Fillunger but for people who are interested in soil mechanics and who perhaps start to be in doubt as to your person or science. One must show such people that in this case a man ventures with his criticism to approach a subject to which he is absolutely not equal."

One recognizes from Rendulic's writings that he had in no way understood Fillunger's fundamental new approach for the description of the mechanical behavior of saturated porous media.

On December 16, 1936, the council of professors held a meeting. The extent to which life at the Technische Hochschule had been effected by Fillunger's attack became evident in Rector Dr. Friedrich Böck's statement:

"Now I come to the final part of my speech, which contains a large number of points, which are without any doubt inseparable and concern an occurrence that, I can say, has burdened us all seriously for 14 days." Then, the Rector listed the aforementioned points chronologically, from the appearance of the pamphlet to the letter from Dr. Fröhlich. Furthermore, he mentioned that von Terzaghi had already expressed the urgent wish to comment on Professor Fillunger's attack in the current meeting. The Rector had convinced von Terzaghi to hold off temporarily and to save his defence for later. In the same meeting, Professor Fillunger had applied for a discussion about von Terzaghi's teaching program. However, the application had been refused by the council against the vote of Professor Fillunger.

After having studied Fillunger's pamphlet very carefully between December 3 and 6, 1936, von Terzaghi and Fröhlich began immediately with the formulation of their reply on Monday, December 7, 1936. Obviously, von Terzaghi did not dare to fight on the theoretical field against Fillunger, who was known in the German-speaking countries as a theoretical scientist with well-founded knowledge in mechanics. He left it to Fröhlich to refute those of Fillunger's attacks which had been made on theoretical ground.

They worked almost every day on their brochure, including Christmas, 1936, finally finishing the manuscript at the very beginning of January, 1937.

In their preface to the rebuttal paper *Erdbaumechanik und Baupraxis: Eine Klarstellung*, von Terzaghi and Fröhlich (1937) pointed out that soil mechanics was a part of applied hydraulics (in a certain sense, a strange statement). Moreover, they emphasized again and again that soil mechanics served to solve practical problems. The preface contained the term "practical" nine times. In no sentence did the term "scientific" appear. It may be that Fillunger's attack had made both authors feel insecure to fight against the theorist Fillunger in the scientific field. Therefore, the authors of the rebuttal restricted themselves mainly to practical problems.

The reply by von Terzaghi and Fröhlich consisted of three main parts: part A treated the history of the creation and the practical applications of soil mechanics; part B contained the rejection of various criticisms on the theory of the settlement of clay layers and, finally, part C contained a brief review of laboratory tests in the field of soil mechanics.

Part A, *Die Erdbaumechanik als technische Wissenschaft*, was written by von Terzaghi. He first repeated the main points of Fillunger's attack. Then he described extensively the creation of modern soil mechanics in Sweden, in the United States, in Germany, and in Turkey, where he had begun his fundamental work in 1917. He reported that the great success of his book from 1925 was based partially on the fact that it contained the solutions to many important practical problems. Later, von Terzaghi gave a brief overview on the historical development of foundation engineering and also discussed the importance of modern soil mechanics.

In the next section of part A, von Terzaghi considered the practical application of soil mechanics. The practical problems of soil mechanics can be summarized as follows:

a) Numerical description of the technically important properties of soil as, e.g., the compression, permeability, shear resistance etc.

b) Avoidance of the role of pure chance during the construction of artificial soil bodies.

c) The setting up of rules for the adjustment of road constructions for artificial roads based on the consistency of the ground.

d) Elaboration of reliable methods for the measurement of motions and pressures on finished buildings in order to gain numerical experience.

e) Revealing the qualitative relations between cause and effect in underground engineering.

f) Elaboration of roughly approximate methods for the numerical prediction of effects in underground engineering.

Subsequently, von Terzaghi discussed points a) through f) in detail, and he refused Fillunger's criticism that an application of the new soil mechanics to civil engineering did not exist and that there would never be such an application.

In the final section of part A, von Terzaghi considered the theories of soil mechanics. These dealt with, according to von Terzaghi, stability of slopes, the earth pressure on retaining walls, the methods of dynamic underground investigations, the stress distribution in loaded foundation soil, and the set-

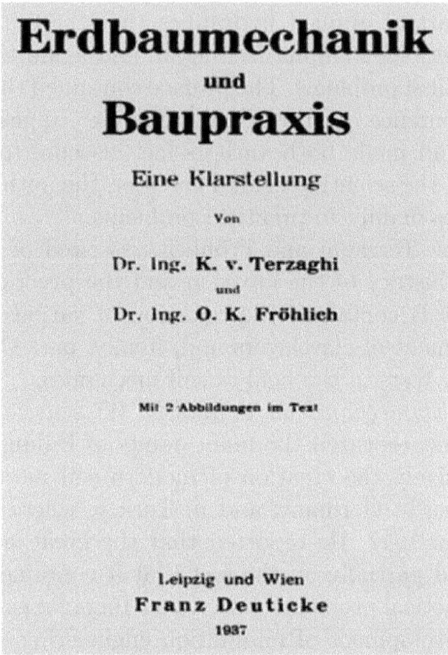

Erdbaumechanik
und
Baupraxis
Eine Klarstellung
Von
Dr. Ing. K. v. Terzaghi
und
Dr. Ing. O. K. Fröhlich
Mit 2 Abbildungen im Text
Leipzig und Wien
Franz Deuticke
1937

Defense brochure by
Karl von Terzaghi and Otto K. Fröhlich

tlement of clay layers. He discussed the settlement of clay layers in detail because only this area had been attacked by Fillunger. Karl von Terzaghi repeated all the basic assumptions of his theory, developed his partial differential equation, and stated that the main predictions of his consolidation theory had been proven again and again by tests and by experience, although some problems had yet to be solved. Nevertheless, there had been a great amount of progress in contrast to the state of the science in 1920.

Part B of the reply *Erdbaumechanik und Baupraxis* was formulated by Fröhlich. He proved that many attacks against the consolidation theory were based on improper statements by Fillunger. In particular, Fröhlich included the acceleration in the derivation of von Terzaghi's differential equation – a main point of Fillunger's approach to the consolidation problem. However, he had in no way understood Fillunger's completely new idea for treating saturated porous bodies. He criticized Fillunger for not having included a constitutive equation for the effective stresses of the porous solid, although Fillunger listed this problem as a desirable generalization of his theory under point 4. Moreover, Fröhlich pointed out that an extensive discussion of Fillunger's theory would be published elsewhere. This did not happen. Instead, Fillunger's theory was completely ignored by the von Terzaghi school in the time to follow.

Fröhlich then refused Fillunger's criticism of the approximate solution of von Terzaghi's differential equation and the theoretical treatment of the test results in the laboratory. He showed that there were some considerable mistakes in Fillunger's attacks. Indeed, this was the case. This was, in particular, evident in Fillunger's criticism of the use of differential calculus on the part of von Terzaghi and Fröhlich. Fillunger had assumed that both authors had applied a special calculus in order to gain the total differential of the pore number, and he spoke of the "Terzaghi-Differential". However, Fillunger made an unimaginable mistake when he introduced the constant value 1 for the variable length l of a clay sample. It is hard to understand how such a conscientious scientist could overlook such a basic fault. It may be that he was blinded by his extraordinary hate for von Terzaghi.

Finally, Fröhlich summarized his view by stating that Fillunger's criticisms were directed against a theory which described the consolidation problem sufficiently for practical purposes. He showed that this criticism was based in part on improper mathematical rules, and that the relations developed by Fillunger to describe the consolidation problem were practically useless because their solution was not known.

Part C, *Versuchstechnik und praktische Anwendungen der Erdbaumechanik*, was again written by von Terzaghi. This part dealt with the objections raised by Fillunger against the test technique and the applications of soil mechanics. Whereas von Terzaghi had described the test technique and practical application of soil mechanics very generally in part A, he now discussed these points in detail in part C, refuting Fillunger's objections.

The reply was finished as already mentioned in the very beginning of January, 1937, and was sent to colleagues, friends, and other individuals in the field of soil mechanics. Enclosed was a handbill: "To whom it may concern: In December, 1936, Dr. Fillunger, o. ö. Professor at the Technische Hochschule of Vienna and member of the *Österreichischen Ingenieur- und Architekten-Vereines* published by himself a pamphlet on soil mechanics. Although Mr. Fillunger, as a person far away from the subject, can not have any insight into the practical requirements of underground engineering and has never carried personal responsibility for the success of a foundation, he believes himself to be competent to give his opinion on soil mechanics as a technical auxiliary science. In order to weaken the practical objections of Mr. Fillunger against soil mechanics, a reference to the numerous valuable practical achievements of this recently developed technical auxiliary science is sufficient. These achievements have been treated again and again in the technical literature. A survey of these can be found in the enclosed brochure *Erdbaumechanik und Baupraxis*: *eine Klarstellung*, formulated for the purpose of clarifying the facts. The brochure also contains a discussion of the mathematical portion of Fillunger's criticism. Although Mr. Fillunger is in his own special field in this part of his criticism, he violates not only the rules but also the spirit of applied mathematics and his criticism unfortunately does not contain even one constructive idea. The publication 'Erdbaumechanik ?' formulated by Mr. Fillunger was sent not only to experts but also to lawyers, editors, administrative officers, institutes and even to the editorial staff of newspapers. It came thereby to numerous persons whose interest in the treated subject was created exclusively by the allegations contained in the publication. In addition, the publication is composed in such a way that it is not only related to my person, but it also heavily jeopardizes the reputation and the economic existence of my former and present co-workers, and, in my opinion, the reputation of the Viennese Technische Hochschule as well.

Finally, it must be emphasized that I have only succeeded in obtaining an extremely incomplete list of the addresses of the individuals who have received Fillunger's brochure. Due to these circumstances, I am forced to direct the request to the receivers of this handbill, that they publicize the contents if possible and that they pass on the enclosed copies of my brochure to experts who are interested in the subject or to those institutions and individuals that have probably received Fillunger's publication 'Erdbaumechanik ?'. The sarcastic and suspicious remarks with which Fillunger's brochure is filled have in no way supported the factual purpose. It is natural that the academic staff of the Hochschule where Fillunger is appointed will deal with this affair. The handbill will be sent to institutes and individuals in all countries where Fillunger, according to his own statement, has distributed his brochure. These countries are: Austria, Germany, Danzig, Switzerland, Czechoslovakia, The Netherlands, Poland, Norway, Italy, Latvia, Russia, Turkey and the United States of America.

Vienna, January, 1937
Dr.-Ing. Karl von Terzaghi
o. ö. Professor at the
Technische Hochschule of Vienna."

Karl von Terzaghi and O.K. Fröhlich received a considerable amount of positive replies. These were appreciative of the purely factual tone of the rebuttal.

The disciplinary committee (Disziplinarkammer) met on December 21, 1936, in the Rectorat of the Technische Hochschule at 10.15 a.m., and not at 6 p.m., as scheduled. Present were:

President:	Magnifizenz Pror. Prof. Ing. List
Members:	o. Prof. Dr. Mache o. Univ. Prof. Dr. Josef Hupka
Alternates:	o. Prof. Dr. Kruppa o. Prof. Dr. Richter
Disciplinary lawyer:	o. Univ. Prof. Dr. Merkl
Secretary:	Dr. Stein

The president, Prof. List, opened the session by welcoming the members and thanked, in particular, the professors of the University. He then read out a letter from Magn. Böck to the disciplinary lawyer, Prof. Dr. Merkl, and Dr. Merkl's response of December 10, 1936, as well as a letter from Magn. Böck to the president of the disciplinary committee containing the disciplinary information against Prof. Dr. Fillunger, a letter from Prof. Dr. von Terzaghi to Magn. Böck with the request to open a disciplinary action against Prof. Dr. Fillunger, Dr. Fröhlich's letter, and a letter written by Prof. Dr. von Terzaghi (without enclosures) to Magn. Böck from December 21, 1936.

The disciplinary lawyer Prof. Dr. Merkl pointed out: "The problem on hand touches the fundamentals of the freedom of science and teaching. The opportunity to criticize must be given. However, the limits of the discipline must be guaranteed. Prof. Dr. Fillunger raises defamatory attacks. Thus, an order to open a disciplinary action is necessary:

1. Prof. Dr. Fillunger must be called as an accused person. He must factually justify his attacks and must say whether personal conflicts existed before this publication.

2. Prof. Dr. von Terzaghi and Dr. Fröhlich are to be called into the witness-stand. They are also to be examined as to whether the pamphlet has personal backgrounds.

The main issue, however, is the need for an expert opinion; this must be written by a commitee of experts. In order to write the expert opinion completely independently, reviewers from outside are to be incorporated, e.g., from other Austrian Technische Hochschulen (or even from abroad). Prof. Dr. Fillunger starts attacks against his colleagues and, moreover, an attack against the scientific image of the staff of professors. These accusations are, however, too vague, so that I will not yet apply for the suspension of Prof. Dr. Fillunger from his duties. The question of suspension will become ripe first after the expert opinion has been finished."

Professor Dr. Hupka was of the opinion that the disciplinary action could be commenced after the decision of the experts.

This opinion was strictly refused by the disciplinary lawyer Prof. Dr. Merkl, who remarked in a severe tone that the opening of the disciplinary action was necessary because the accusation had become widely public and therefore the involved groups would be able to see that the Hochschule was already officially examining the matter. The expert opinion would only be able to state the gravity of the offense.

Professor Dr. Hupka replied: "Without doubt, the opening of the disciplinary investigation should be voted. However, within the framework of the disciplinary action, expert opinion must be brought in by the professors' council."

In contrast to this professor's view, Dr. Merkl answered: "The expert opinion must be objective; therefore, no one depending on Prof. Dr. Terzaghi or one of his disciples, may be a member of the expert committee."

Then, the president, Prorector Prof. Dr. List, recommended that the Rector should nominate the experts. Moreover, he stated that the disciplinary lawyer should request the opening of the disciplinary investigation.

Professor Dr. Merkl formulated his motion: "Making the accusation concrete, by citing the corresponding parts of the brochure 'Erdbaumechanik?' of Prof. Dr. Fillunger, has to take place first in the resolution of reference. I therefore move according to §113 D.P.[49] to pass the resolution of the opening of a disciplinary investigation against Professor Dr. Paul Fillunger due to a suspicion of offense of regulations founded by the publication and distribution of the printed book 'Erdbaumechanik?'."

In the ensuing discussion, Professor Dr. Kruppa declared that members of the expert committee should also be nominated by the professors, Dr. Terzaghi and Dr. Fillunger. Professor Dr. Hupka remarked that the expert committee should be able to examine both sides. Prorector Prof. Dr. List proposed that the Rector set up a mixed commission, and that this commission determine further experts who would then be delegated to an expert committee.

Finally, the president Prorector Prof. Dr. List put the motion (of the disciplinary lawyer) concerning the opening of a disciplinary investigation to a vote. The motion was universally accepted. The president stated: "Professor Dr. Fillunger is now to be examined as an accused person; he has to prove the truth of his accusations. Moreover, it must be proven whether there are personal motives. In addition, an expert opinion must be brought in. The composition of the expert committee should be arranged by the Rector on the basis of the nominations by the restricted commission, after the hearing of the accused and attacked persons. The task of the expert committee will be the special scientific review of the technical explanations in Professor Dr. Fillunger's brochure."

In addition, Professor Dr. Merkl pointed out that the decision to open a disciplinary investigation had to be served to Professor Dr. Fillunger and accompanied by information on his right of appeal. The hearing ended at 11.15 a.m.

The Rector of the Technische Hochschule of Vienna and Professor Dr. Fillunger were informed of the decisions of the disciplinary senate on the same day. Professor Dr. Fillunger was simultaneously informed that no right of appeal was admissable. The Rector appointed the Administrationsrat Regierungsrat Dr. Josef Goldberg on the same day to be the investigation-commissioner in the disciplinary investigation against Professor Dr. Paul Fillunger.

In addition, the Rector appointed the professors, Dr. Franz Jung (mechanics), Dr. Ludwig Flamm (physics), Dr. Lothar Schrutka-Rechtenstamm (mathematics), and Dr. Josef Kozeny (hydraulics) to form a commission. He

[49] Dr. Merkl referred to the federal law regarding disciplinary action against professors.

requested that the commission be formed by Monday, January 4, 1937, and entrusted the commission with the task of nominating experts for the investigation committee. This committee was set up on January 7, 1937, and consisted of the professors, Dekan Dr. Karl Wolf (mechanics), Dr. Rudolf Saliger (concrete construction), Dr. Friedrich Schaffernak (hydraulic engineering), Dr. Franz Jung (mechanics), Dr. Ludwig Flamm[50](physics), Dr. Schrutka-Rechtenstamm (mathematics), Dr. Franz Aigner (physics), Dr. Alfred Lechner (mechanics), and Dr. Josef Kozeny (hydraulics).

The Rector remarked in his letter that, in the case of inviting an out-of-town expert, the *Rectorat* had to be informed for the reason of covering the expenses. Moreover, he made an appeal to begin immediately with the investigations and to finish them, if possible, by January 23, 1937.

Referring to statements of von Terzaghi, the professors, Dr. Jung, Dr. Lechner, Dr. Wolf, and Dr. Flamm were recommended by Prof. Dr. Fillunger, whereas the professors, Dr. Schaffernak, Dr. Aigner, and Dr. Schrutka-Rechtenstamm were recommended by von Terzaghi. Von Terzaghi ignored Prof. Dr. Salinger because of factual differences. Concerning recommendation of Prof. Dr. Flamm By Prof. Dr. Fillunger, Prof. Dr. von Terzaghi did not tell the truth. In a letter of December 28, 1936, von Terzaghi requested Professor Dr. Flamm to act in the commission; in his hand-written reply of December 30, 1936, Professor Flamm agreed to von Terzaghi's request.

The appointed investigation-commissioner, Dr. Goldberg, began immediately with his questioning. He started questioning Professor Fillunger just one day after his appointment, on December 22, 1936. First, Fillunger discussed in detail the whole previous history of the affair, in particular the dispute over the uplift problem. Concerning his first dispute with von Terzaghi about the uplift problem, Fillunger had a completely different view of the affair than von Terzaghi had. Dr. Goldberg stated in the questioning protocol: "A close involvement with the scientific work of Prof. Terzaghi happened first when Prof. Fillunger in his function as dean in the academic year 1932/33 had to intervene in vivae voce[51]. Shortly after the viva voce of Leo Casagrande

[50] Ludwig Flamm, born on January 29, 1885, came from a clock-maker family in Vienna. Here,he graduated from the university. In 1909 he passed the viva voce at the University of Vienna, became an assistant and did his habilitation thesis in 1916. In 1922, he was offered a full professorship at the Technische Hochschule in Vienna where he was employed until 1956, when he retired.

Flamm's contributions to theoretical physics lay in different fields. His paper *Contributions to Einstein's gravity theory* is probably the most valuable one. In 1929 – 1931 he was head of department and in 1950/51 Rector.

Professor Dr. phil. Ludwig Flamm died on December 4, 1964.

[51] Among other things he was involved later in the doctor thesis of Arthur Casagrande, who had written his doctor thesis not completely from scratch but had sent in some papers on the structure of clay and had relied on von Terzaghi's and Schaffernack's goodwill. Fillunger remarked in the questioning of December 22, 1936, that Arthur Casagrande had spoken of the so-called honeycomb structure of clay which could not be proved in Fillunger's opinion.

(brother of Arthur Casagrande) and von Terzaghi's assistant, he gave the impetus for writing a scientific paper to the young doctor, in which he could make use of his dissertation and should touch other respective problems which were connected with single questions concerning the uplift in barrages in the papers of Hoffman, Fillunger and Toelke published in the years 1928 – 1932; Prof. Dr. Fillunger did not want to pick them up again by himself. For this purpose Prof. Dr. Fillunger gave Leo Casagrande a respective publication of O. Hoffman, his own rebuttal: *Uplift and Underpressure in Barrages*, a subsequent polemic with Hoffman and a paper of Toelke in the *Ingenieurarchiv* as documents ... Leo Casagrande informed Prof. Dr. Fillunger that he had had a consultation about this with Prof. Dr. Terzaghi, who stated that Prof. Dr. Fillunger was done injustice in the literature and he would take charge of the matter; Prof. Dr. Fillunger took notice of this in the presence of Leo Casagrande. Approximately two to four weeks later Prof. Dr. Terzaghi visited the accused and declared to him that something was not correct in the calculation of Prof. Dr. Fillunger." Both professors discussed the uplift problem several times before the meeting in the café Kuhnhof already mentioned. "During this occasion Prof. Dr. Fillunger explained to Prof. Dr. Terzaghi, he requested or entreated him, to let his *attempts* be and not to publish the manuscript, because otherwise Dr. Fillunger must come to speak of everything in the polemics which could come to a "ripping apart".

In the following paragraphs Fillunger's view of his difficulties to get his paper published were considered where he suspected von Terzaghi of trying to prevent the publications. Moreover, Fillunger pointed out that two papers by himself were reviewed by Dr. Rendulic, von Terzaghi's co-worker, in the Zentralblatt für Mechanik in a very brief manner and that von Terzaghi's publications were discussed extensively. Fillunger concluded that in the coverage of this journal von Terzaghi's works were preferred.

In the inquiry Fillunger presented also a letter of O.K. Fröhlich from May 3, 1934, in which Fröhlich admitted the heavy doubts concerning the capillarity-pressure, treated in the book *Erdbaumechanik* in 1925, and in which he added further doubts on results in von Terzaghi's book. He denoted, finally, the so-called theory of the *hydrodynamic stress phenomena* as the only good thing in the book *Erdbaumechanik*. Fillunger drew the conclusion that even in the closest circle of co-workers of Prof. von Terzaghi, the drawbacks in *Erdbaumechanik* were well-known. However, they did not reach the technical public.

Furthermore, Fillunger stated, as a result of his bad experiences in attempting to get his paper published, that the only possibility to reach the public was to edit his brochure by himself and that it was necessary to use a very sharp tone in his pamphlet to shake a broad public into action. He added that he wanted to give Prof. von Terzaghi the opportunity to hand in an honorary action for defamation in order to discuss the matter scientifically in the public.

In the questionings of February 17, 18, and 19, 1937, Prof. Fillunger was examined about the personal defamations in the pamphlet *Erdbaumechanik?*. Fillunger's answers were, in general, factual and exhaustive. In the course of his examination Fillunger attacked sharply the thermodynamic parable. He pointed out that this parable should be correctly called parable of heat propagation. Obviously, this differential equation of heat propagation served as a model which was violently imitated for the formulation of the differential equation of the porewater flow. "Only if the differential equation of the porewater flow would have been developed independently of the parable of the heat propagation and the analogy with the parable of the heat propagation would have come out later, then one can speak of a parable ... The mystification of the reader exists in the fact that the laws, found in the range of heat propagation, were considered as a confirmation for the laws of the porewater flow discovered seemingly independently."

After these remarks about the parable he listed some other examples of mystification in Karl von Terzaghi's work and he finished up with the statement that the main motive for his sharp criticism had been the fact that the theories of von Terzaghi and Fröhlich were liable to cause great damage:

1. In the practice, due to an unjustified confidence in these theories, it could happen that false foundations could be constructed which would weaken the reputation of the engineers. This would be very critical in view of the political situation (Bolshevism).

2. In the teaching program of the Technische Hochschule the theories are taught in the higher semesters where the students do not have relations to the theoretical subjects and therefore can not reflect critically on the theories.

3. In scientific circles it can happen that real science, complicated by great intellectuals, will be refused, whereas superficial simple statements spread rapidly.

The question as to whether he had violated the reputation of the Technische Hochschule by his personal attacks was denied by Fillunger.

Fillunger could not sign the protocol because the fair copy was typed after the sad affair.

The questioning of von Terzaghi began on December 23, 1936, from 9.30 a.m. to 12.30 p.m. The topics were the same as in the questioning of Fillunger, namely the previous history of the affair (the polemics in connection with the uplift problem), some factual points in the brochure *Erdbaumechanik ?*, and the personal defamations in Fillunger's pamphlet. Concerning the uplift affair Karl von Terzaghi repeated his view that he got involved in the uplift problem in the fall of 1932 when he studied Fillunger's paper on uplift developed in the years, 1913 – 1930. Moreover, Karl von Terzaghi summarized Fillunger's factual accusation in *Erdbaumechanik ?*: "The factual part of Fillunger's attack culminated in the statement that the theory of the gradual consolidation of loaded clay layers, developed by the witness, is absurd and that building-technical applications of earthwork mechanics neither exist nor can exist. The earthwork mechanical research is superfluous because the experience of the real practical men is sufficient for mastering civil engineering underground construction problems.

The theory, developed by the witness and denoted as absurd by the accused, raises the claim to deliver the approximate description of physical proceedings which are represented by the ödometer test in the laboratory. A theory of this kind can only be denoted as absurd, if no similarity between calculation results and the reality to be described, exists ... In the course of the last ten years the ödometer test has been generally adopted. It has been repeated countless times and there has been no single case known which has given cause for a doubt on the practical usability of the theory. Besides, hitherto nobody has succeeded in developing a theory which demonstrably comes closer to reality than that of the witness, the accused included."

Finally Prof. Dr. von Terzaghi pointed out that Prof. Dr. Fillunger, in connection with his factual accusations, had always repeatedly indicated that the witness propagated the *false doctrine* of earthwork mechanics for the purpose of self-enrichment. "For the proof to the contrary of this accusation the witness renders with emphasis that he carries on his business as a reviewer only to such extent as this advances the purpose of his research and that he has repeatedly renounced royalties totally or partly, if he could get the promise of carrying out scientific value observations."

Further questioning was performed on February 10, 1937, and March 3 to 5, 1937. In these last hearings, von Terzaghi was confronted with some factual accusations and, in particular, with the personal defamations in *Erdbaumechanik ?*, and he listed the most outrageous points. Moreover, he gave an interesting inside-look into his method of scientific working. He pointed out that, particularly in the first decade of his research, his method was determined by the fact that his field had in no way been scientifically investigated. In such a situation, the success of his efforts would depend exclusively on a talent which he called *physical scent* and which could not be learned. His method of working could be described as follows: as soon as a phenomenon in nature excited his interest, he tried at first to understand purely intuitively the progress of the phenomenon due to the properties of the substances which were involved in the progress. He therefore repeated the experiment again and again over longer time frames. When, in 1923, he made efforts to describe analytically the nature of the settlement of clay layers, he gave up the attempt in an exhausted state after several weeks of hard work. Three months later, he succeeded easily within two hours, and after five further hours, the derivation of the fundamental equation of the consolidation problem was completed. This method is sometimes called intuition. The next step consisted of the mathematical description of the conception of the nature of the problem, which he had already formed in his mind, in which he neglected apriori all such factors which he considered intuitively as unimportant. Thereafter, he invented such experiments whose results allowed for the proof of the theoretical ideas. The completion of the investigation consisted of performing the experiments. If, then, the experiment confirmed the correctness of the basic idea, the further development of the theory could be performed with the help of recognized and generally used calculation rules. Because he did not have, however, any interest in this part of the work, he encouraged qualified colleagues to continue the work. The result of such an encouragement becomes evident in the book *Theorie der Setzung von Tonschichten*, in which Fröhlich did the strenuous work of the evaluation of the partial differential equation. In this connection, von Terzaghi declared that he had never read and would never read the sections worked out by Fröhlich, since he was satisfied if he knew the assumptions and the results. The role of the *physical scent* becomes very clear in the fact that most of his experiments were not new. The difference between his own work and that of his predecessors consists only in the manner of observation and in the apparent slight changes of the test arrangements. In order to discover a new relation through an experiment, one must already suppose its existence beforehand, and must perform the test-program in such a way that the validity of the supposition becomes evident in the test results. Von Terzaghi became convinced, over the course of several years, that the talent for doing this kind of work could not be taught. Despite severe efforts, he succeeded only occasionally in discovering, among his students and co-workers, young people who had at least a point of attachment to this ability. More widespread, even in mathematically trained circles, are such minds that cannot understand a physical process even if the description is already completed. According to the aforementioned way of working, von Terzaghi started almost from a more or less intuitively gained set of conceptions; he proved, later on, whether these conceptions were correct. In his papers, however, he had to choose the inverse manner. From this inversion of his thinking process, there resulted not infrequently inconsistent jumping from one idea to another.

With this glimpse inside von Terzaghi's working-style, the questioning on March 5, 1937, was closed.

Not only Fillunger and von Terzaghi were questioned by Dr. Goldberg in 1936 and 1937, but also Dr. Fröhlich and other professors of the Technische Hochschule of Vienna. As early as December 23, 1936, the questioning of Fröhlich concerning his relations to von Terzaghi and Fillunger had started, lasting from 12.30 p.m. to 1.25 p.m. A further hearing with Fröhlich took place on February 24, 1937. He expressed his admiration for von Terzaghi's talent for intuitively recognizing important physical phenomena. Fröhlich recalled a related statement of the famous hydraulician Professor Forchheimer: "Professor von Terzaghi has a distinct ingenius talent for technical-physical problems." Furthermore, Fröhlich pointed out that von Terzaghi's papers contained, in general, only the main ideas, so that it was not surprising that they had not been understood for a long time.

Concerning his acquaintance with Prof. Dr. Fillunger, Dr. Fröhlich remarked that he had met him at the airfield in Aspern in 1918 and that his relationship to him had always remained that of a comrade. Moreover he stated that he had – on the occasion of a visit of Prof. Dr. Fillunger in May 1934 – aspired to abolish the differences in the opinions about the effective surface porosity between Prof. Dr. von Terzaghi and Prof. Dr. Fillunger.

Professors of the Technische Hochschule of Vienna were questioned in January, February, and March, 1937, mainly concerning the jeopardization of the reputation of the Technische Hochschule by Fillunger's attack.

The commission to set up an investigation committee had already met in December, 1936. On Tuesday, December 29, 1936, Professor Flamm declared that one should not construct a new theory but should keep the model of von Terzaghi. Moreover, he doubted the validity of the incompressibility condition for the partial water body.

As already mentioned, the investigation committee began its work at the very beginning of January, 1937. It held eighteen meetings. The leading member of the committee was Professor Flamm, a son-in-law of the famous physicist Professor Ludwig Boltzmann. He formed a strong group with the professors Schaffernak (hydraulics, and a friend of von Terzaghi) and Aigner (physicist), who had no opinion of his own in most cases, but agreed with Flamm and Schaffernak. There was obviously some animosity between Flamm and the professors of mechanics and, in a meeting in January, 1937, Professor Flamm asked his colleagues whether they had ever heard anything about mechanics. Upon hearing this, Professor Jung severely denounced Flamm's arrogant tone. The professors of mechanics investigated the scientific side of the affair very carefully. In particular, Professor Lechner, together with his co-worker Dr. Heinrich, showed that von Terzaghi's partial differential could be gained from Fillunger's basic system of differential equations, using, however, strongly simplified assumptions. Lechner's and Heinrich's investigations later became part of the body of expert opinion. They were also later published by Heinrich (1938). For Flamm and Schaffernak, the investigations of the mechanics-group were obviously a little too boring. In order to convince his colleagues that von Terzaghi's work was sound, at the end of January Schaffernak constructed a simple mechanical model of the consolidation problem. However, the members of the mechanics group continued their work to develop von Terzaghi's differential equations based on Fillunger's fundamental equations.

A decisive meeting of the investigation committee was held on January 30, 1937, between 11 a.m. and 1 p.m. During the discussions, it became evident that Fillunger had made serious mistakes in Sections III and IV of his brochure *Erdbaumechanik?*. This statement was devastating for Fillunger's position. The committee also visited von Terzaghi's laboratory, and the members saw for themselves that von Terzaghi and his staff worked with scientific methods.

Obviously, Professor Salinger reported the results of the meeting to Fillunger in the afternoon, to which Fillunger expressed his feelings in a very loud voice, leading to a gathering of the students outside in the hall.

In a letter of February 5, 1937, addressed to Professor Wolf, the chairman of the investigation committee, Fillunger requested that the committee allow him to present a statement to the reply *Erdbaumechanik und Baupraxis: Eine Klarstellung* by von Terzaghi and Fröhlich. In a letter of February 8, 1937, Professor Wolf answered that the members were only prepared to receive Fillunger's statement privately, as they had done with von Terzaghi's and Fröhlich's defending report.

Fillunger finished his statement in a great hurry, and in a short letter to the members of the committee he apologized for possible mistakes in his statement, *Vorläufige Erwiderung auf die Verteidigungsschrift der Herren v. Terzaghi und Fröhlich*, because of his haste. In the very beginning, he repeated his main point of criticism:

"The 'Theorie der Setzung von Tonschichten' by Mr. v. Terzaghi and Mr. Fröhlich as well as the right to completely reject this theory from a scientific point of view by the author of this reply, *stand and fall* with the answer of the question as to whether the differential equation of the porewater-flow can be scientifically founded, as it has been developed by v. Terzaghi and apart from this has been supported by him and Mr. Dr. Fröhlich. Therefore, this question should be discussed here first of all. Compared to this question, other questions, for example, those referring to the experimental basis, to the practical application and to the representation form in the book 'Theorie der Setzung von Tonschichten', are unimportant ..."

In the first part of Fillunger's treatise, the theoretical part, he remarked that the formal correspondence of von Terzaghi's differential equation with the Fourier-equation for non-stationary heat propagation could not be sufficient in scientifically founding the differential equation. He then discussed in detail all terms in the derivation of von Terzaghi's differential equation and criticized the analogous conclusion, as well as the attempt on the part of Dr. Fröhlich to include the acceleration in the differential equation. He also tried to prove that Dr. Fröhlich's approximate solution was false. In the second and third section of his reply, Fillunger treated the consolidometer test and justified his criticism.

After a very intense committee meeting on February 13, 1937, Professor Schaffernak became ill and was not able to attend the meetings in the following weeks. Thus, von Terzaghi lost a personal friend on the investigation committee. However, the affair was already decided, and the subsequent meetings served only to formulate the details of the result.

Karl von Terzaghi was deeply hurt by the whole affair. There were also at the time several ugly rumors about his role in connection with the construction of the Reichsbrücke in Vienna. In 1935, the news had already been spread in Germany, by a former co-worker of von Terzaghi, that his position as a specialist in Austria had been severely shaken. Among the teaching staff of the Technische Hochschule of Prague, it was an open secret that von Terzaghi had at first proposed a suspension bridge in Vienna, and thereafter recognized that the soil was not suitable for the anchoring of such a construction. In Vienna, the newspapers reported that the high costs of the Reichsbrücke were a result of the excessive caution of von Terzaghi. Although the responsible persons in the government declared that the news had been invented by reporters, it was clear to von Terzaghi that the information had come from certain circles in the government. von Terzaghi also had some "enemies" in Berlin.

After Fillunger's attack, and the rumors cited above, von Terzaghi believed that there was a "Hintermann" (a man operating behind the scene). Obviously, he suspected his colleague Professor Dr. Saliger (professor for concrete construction) as a possible source of such talk. At the very beginning of January, 1937, von Terzaghi called Saliger by phone and then subsequently met with him on January 8, 1937. Saliger promised to send von Terzaghi a letter. This was done on January 9, 1937. In the letter, Saliger assured von Terzaghi that he himself had never spread rumors about him and that he was upset by Fillunger's attack.

On January 11, 1937, the dean of the faculty, Professor Wolf, informed the Rector that von Terzaghi had cancelled his lectures. An urgent request from the Rector convinced von Terzaghi to take up his lectures again.

As has already been pointed out, von Terzaghi and Fröhlich worked hard to fend off Fillunger's attack, along with certain rumors. After completing the reply (von Terzaghi and Fröhlich, 1937), von Terzaghi worked on the opinion paper of the Reichsbrücke, and met with some members of the committee. In February, 1937, von Terzaghi formulated a memorandum, consisting of three parts. The first part explained extensively the previous history and the factual background behind the attack. He repeated the story of the disagreement over the uplift problem, and formulated a first statement regarding the factual side of Fillunger's attack. The essential points had already been included in the reply from von Terzaghi and Fröhlich (1937). In the appendix of his memorandum (part one), von Terzaghi denied the accusation that he had spread his view on soil mechanics solely for profit-seeking reasons. He stated that his consulting work was only dedicated to those technical problems which were interesting from a scientific point of view. Moreover, he declared that his financial gains had always been very modest, and he proved this by various examples.

In the second part of his memorandum, von Terzaghi gave some advice to the members of the investigation committee, which consisted mostly of mathematicians, physicists, and mechanicians. He defined soil mechanics as a branch of applied hydraulics, and stated that soil mechanics served exclusively for practical purposes. He further stated that the essential and seriously offensive remarks by Fillunger could be judged only by representatives from the applied fields or by experts in civil engineering. The task of the mathematicians, physicists, and mechanicians could only consist in discovering the errors in Fillunger's reasoning, which he must have made in order to declare a theory to be absurd which had been proven in the laboratory as well as on the building site. Then, von Terzaghi suggested a detailed schedule to the committee in order to accelerate the passing of a resolution. Von Terzaghi was very much interested in the immediate clarification of the attack. Besides the detailed schedule, he also developed a questionnaire to be answered by the members of the committee, via the multiple-choice-method. Obviously, this memorandum never came to the knowledge of the committee; this was certainly good for von Terzaghi's position. One can imagine how the members of the committee, all professors in their own right with their own views on the whole affair, would have reacted to such a strong attempt to influence their opinion.

In the third part of his memorandum, von Terzaghi presented a broad exposé of Fillunger's scientific talent. First, he discussed the papers of Fillunger from 1912 on. He accepted the papers on the elasticity theory, but rejected the papers which dealt with new physical phenomena, for example, with the uplift problem. He heavily criticized Fillunger's first paper on the uplift problem, published in 1913, and the subsequent papers, as well as Fillunger's polemics against Hoffman and Lehr. Von Terzaghi stated that Fillunger had a particular talent for reacting when he recognized that he was in error. In this case, he would mask his results again and again by going far back into his mathematical considerations and drawing the readers' attention away from the weakest point of his theory. Moreover, he would invent new mathematical proofs in order to show that his formulas were correct, and would present this for the most part in a rather confused manner. On the basis of all these experiences, von Terzaghi finally gained the following impression of Fillunger's intellectual personality: "Fillunger connects good mathematical training with a typical defect in physical talent, and with a distinct defect in scientific self-criticism. He does not have any imagination of the possible differences between the physical reality and the abstract assumptions on which the mathematical description of the problem is based. With this, the apparent contrast between Fillunger's success in the field of the elasticity theory and his failure considering physical problems is explained. If the starting point for a thinking process is uniquely given as, for example, in the elasticity theory, every mathematically trained person is able to calculate further in a correct way. Fillunger was in this position when he developed his theoretical papers in the fields of the elasticity theory and the strength of materials. If, however, the formulation of the approach requires physical understanding, he moves away from the reality caused by his purely abstract ideas, and does not return to the reality. This was the case in the development of his uplift theory and was repeated in his criticism of the consolidation theory. Moreover, if Fillunger does not understand the ideas of another researcher, he declares them as false and then invents, in the purely abstract field, complicated proofs for his unfavorable judgement and he does not change his view, notwithstanding the most obvious experimental proof to the contrary. In the course of the last decade, these described oddities of Fillunger's scientific activities have raised another point, namely the emphasis of his own importance by mockery and disparagement of persons who express other views. This point is the source of his attacks against other scientists 'to free science from harmful elements'."

However, Karl von Terzaghi's review of Paul Fillunger's achievements concerning the uplift and effective stress phenomena failed in certain parts completely and his remarks on these problems are confusing and ridiculous, revealing that he in no way had fully understood these important effects in loaded saturated porous bodies, although he had spent so much time investigating these phenomena: "The uplift depends ... among other things on the microstructure of the material, so that each attempt of a purely theoretical treatment delivers at the best a rough approximate solution, which needs an experimental-technical verification." Moreover he stated: "The uplift has according to our experience the maximal value, which it can generally reach, in pure sand, which is also a homogenous porous body. Therefore it is obvious for every impartial human being that the uplift can take every value between zero and the maximal value, if the microstructure of the homogenous porous body is unknown." It is amazing that von Terzaghi did not recognize that according to Archimedes' principle the full uplift is acting and is, of course, independent of the microstructure and of any strength of material effects, but only on the size of the porosity of the porous body. Furthermore, Karl von Terzaghi did not recognize or did not want to recognize that Fillunger had described the effective stress concept already in 1913 and 1914. His review of Fillunger's (1915) test results concerning the effective stress concept are in no way understandable. He mixed up this concept with uplift; it seems that he had not read Fillunger's valuable paper or the real meaning of uplift and effective stress was not clear in his mind: "In 1915 Fillunger published the results of experimental investigations on test bodies ... The observation results appear to me incontestable. The interpretation of the observed phenomena is, however, in my opinion insufficient, because he (Fillunger, the author) missed the contradiction between his statement on the uplift in cement stone and the test results. Fillunger's inability to recognize the existence of this contradiction originates from the same peculiar intellectual feature as his uncomprehending attitude towards my theory of the settlement of clay layers." One can state that Karl von Terzaghi missed in this point, because of his lack of knowledge of the uplift and effective stress phenomena, the right view of Fillunger's achievements. At least Fillunger had fully understood the effective stress concept and he applied it so masterfully later (Fillunger, 1936) whereas his adherence to his incorrect uplift formula and his explanation of the uplift effect seem to be founded in his stubbornness.

Karl von Terzaghi had obviously a good relationship to Dr. Goldberg, for he informed him of the atmosphere in the council of the professors and allowed him to inspect the questioning protocols, a somewhat strange action, because von Terzaghi derived, without doubt, an important advantage from this privilege against his opponent.

While von Terzaghi's life during the affair, from December, 1936, through February, 1937, can easily be followed by reading von Terzaghi's diary, little is known regarding the course of Fillunger's life during this time. Of course, his daily life was also filled with efforts to defend his polemical pamphlet and further attacks. Already in the council of professors, as has been mentioned, he had applied for a discussion over von Terzaghi's teaching program. This request was, however, refused. Also, in the questioning, he heavily defended his own polemical statements and attacks on von Terzaghi's scientific work. After the appearance of the defending treatise by von Terzaghi and Fröhlich (1937), Fillunger worked intensively on the brochure in question. The Dean Professor Wolf, chairman of the investigation committee, told Fröhlich that Fillunger had, by January 21, 1937, realized his fault concerning his attacks against the mathematical treatment ("Terzaghi-differential" and approximative procedure) adopted in the book by von Terzaghi and Fröhlich (1936). Moreover, Wolf reported that Fillunger had been totally bewildered. After the decisive meeting of the investigation committee on January 30, 1937, Fillunger lost his self-control. In the evening, Fillunger formulated a letter to the Rector which, however, he did not mail:

"January 30, 1937

Euer Magnifizenz!

The paper by Dr. Fröhlich taught me that my offensive remarks were unfounded. I beg all those whom I have hurt to pardon me.

Fillunger"

On the same evening, he destroyed all those letters which supported him in the affair.

During the next weeks, Fillunger nevertheless worked on the aforementioned preliminary reply to the defending treatise by von Terzaghi and Fröhlich. After he had handed it over to the members of the investigation committee, he revised the manuscript in February. Obviously, he was gradually realizing that he had been wrong in some parts of his polemical pamphlet *Erdbaumechanik ?*. The strain on his nerves became too much and he wasted away. Gradually, he came to believe that the only means of escaping the affair lay in committing suicide. He would otherwise probably be suspended from his duty at the Technische Hochschule. However, there were obviously days when he believed he could win the fight, and on those days he worked hard on his preleminary statement, changing some parts completely.

On Saturday, March 6, 1937, Fillunger worked in the Technische Hochschule, as usual. At 11 a.m., he phoned his wife, Margarete, at their apartment in the Messerschmidtgasse 28 in Vienna. One can only guess that someone had informed him that the investigation committee had come to a final decision, voting to support von Terzaghi on all points. After the telephone call, Mrs. Fillunger rushed out of the apartment, completely dazed, returning only after a while.

Around noon, Fillunger came home from the Technische Hochschule, changed his clothes, and left the apartment again at 2 p.m., in great agitation. His wife followed some minutes later, herself also very excited, shouting: "I must hold him back."

In the late afternoon, the couple came back to their apartment in the quaint residential area. Mrs. Fillunger had shared all of the professor's scientific work and was also intensively involved in the polemic conflict. Both made the grave decision to commit suicide.

They prepared the suicide very carefully. Mrs. Fillunger cancelled the daily milk delivery from the shop. This was not surprising, as the couple had often spent weekends outside Vienna. Moreover, Mrs. Fillunger deposited the key for the apartment with the caretaker. That evening, they wrote ten farewell letters and, in addition, Professor Fillunger wrote his last will and testament. Among others, there were farewell letters to the Rector, to his co-worker Dr. Ježek, and to the police. In the farewell letter to the Rector, he included a letter which he had already written on January 30, 1937, and which has been cited earlier:

" 6. III. 1937

The again awakened hope has disappeared recently, more and more I am struck by the sense of my guilt, which I might be able to weaken only by the assurance that I had already destroyed all letters which seem to agree with me on January 30."

In the farewell letter to his co-worker of many years, Dr. Ježek, he thanked him for his cooperation:

"Dear Doctor Jeẑek!

Do not think too unfairly of your superior of many years, to whom you were always a true co-worker. I thank you for all your kindness

Fillunger.

Unfortunately, I was deaf to your warning."

The letters to the police by both Paul and Margarete Fillunger are particularly deeply moving:

"My blindness has left me – the good faith, in which I believed myself to be in not too long ago, does not exist. Heavy attacks upon another person require penance which I can only give to myself. My extremely brave wife will not let me expiate alone.

We ask for the heart[52]-stab before the funeral."

"The great devotion and love for science which my husband always held as his most precious possession, led us, due to a scientific error, into death. Our

[52] It was sometimes customary in Austria to stab dead people into the heart in order to ensure that they are really dead.

honor can only be saved in this way, and the belief be credited to us that only the noblest idea and the most honest strict conviction to serve the truth justified this scientific step."

Mr. and Mrs. Fillunger put the letters on the desk in Professor Fillunger's study room. They put their door bell out of use by winding fabric around it and prepared their bathroom. They brought in an arm-chair and sealed the bathroom door with newspapers. In addition, they fixed a warning outside on the bathroom door: "Bitte kein Licht machen!" ("Please do not turn on the light!"). Both changed their clothes, dressing themselves totally in black. They took soporifics and opened the gas tap.

On Monday, March 8, 1937, at 8:30 a.m., a carrier called to deliver pictures which had been sent by Fillunger's son, Erwin, who lived in Munich. Because no one opened the housedoor in the Messerschmidtgasse 28, the caretaker, who had the key, let the carrier into the apartment. The caretaker, who was very well acquainted with Fillunger's habits, noticed at once that the newspaper was lying unread at the door in the same shape as when she had put it into the door slot. Moreover, the door to one of the four rooms was open and on the desk lay, in great disorder, numerous galley proofs, money and several letters. The paralysing silence was ominous, and when the caretaker and her accompanier saw the sealed bathroom door, they had a hunch that dreadful things must have happened in the apartment. When they courageously opened the door they saw the dead Margarete Fillunger sitting in the arm-chair, and a lifeless Paul Fillunger lying collapsed on the floor. He had obviously fallen from his stool, and he lay with his head close to the feet of his wife.

The suicides of Margarete and Paul Fillunger were a great scandal in Vienna. All of the newspapers in Vienna and environs reported on this event during the following days, describing the facts and searching for the motives. Professor Karl von Terzaghi was upset when he was informed of the suicides, as he declared in a Vienna newspaper, however, first, after a local editor had asked him to say the word *upset* in order to bring a human note into the news.

Nearly all of the members of the Technische Hochschule of Vienna were in a state of shock because of the tragic end of the strange affair, and they were unable to understand the deadly outcome. After the suicide, the Technische Hochschule gave a statement to the press on March 12, 1937, wherein the facts concerning the history of the affair were explained and commented upon. In particular, it was cited from the – then as yet – unpublished expert opinion paper that the factual attacks of Professor Fillunger had had no qualified basis. The statement to the press was strongly influenced by Karl von Terzaghi, who was indignant at the delay of the delivery of such a statement to the press. In order to accelerate the procedure he visited the investigation-commissioner Dr. Goldberg and formulated with him the statement of the Hochschule, which was handed over to the press.

1. Auflage

V. b. b.

Bundesländer 12 Groschen

Das Echo

Zehn Groschen

Nr. 55 Montag, 8. März 1937 Jahrgang 4

Wissenschaftlicher Streit als Selbstmordmotiv

Die Tragödie des Hochschulprofessors Fillunger

Wie der „Telegraf am Mittag" kurz berichtete, wurde heute morgen der Selbstmord des Professors an der Technischen Hochschule, Dr. Ingenieur Baurat Paul Fillunger, entdeckt. Professor Fillunger war gemeinsam mit seiner Gattin Margarete in den Tod gegangen, weil er einsehen mußte, daß er in einem wissenschaftlichen Streit mit einem Fachkollegen Unrecht gehabt hatte.

Professor Fillunger und seine Gattin vergiftet aufgefunden

Der Selbstmord dürfte bereits in der Nacht von Samstag auf Sonntag geschehen sein. Man beobachtete, wie Professor Fillunger Samstag mittags in höchster Aufregung von der Technischen Hochschule heimkehrte. Er dürfte an diesem Tage von der Einleitung des Disziplinarverfahrens gegen ihn erfahren haben. Kurze Zeit darauf eilte Professor Fillunger aus dem Haus, wenige Minuten später folgte ihm in höchster Aufregung seine Gattin.

Als man sie fragen wollte, was geschehen sei, stieß sie nur hervor, „ich muß meinen Mann abholen". Im Laufe des

zehn Abschiedsbriefe hinterlassen, die an seinen Sohn, an Freunde und Berufskollegen gerichtet sind. Diese Briefe werden verschlossen den Adressaten übermittelt werden.

Auf einem offenbar für die Polizei bestimmten Zettel gab Professor Fillunger in großen Umrissen das Motiv seiner außergewöhnlichen Tat bekannt. Er beteuerte auch, daß er und seine Frau „in Gerechtigkeit" gestorben seien und bat, an ihm einen Herzstich vorzunehmen.

Die Streitschrift gegen Professor Terzaghi

Ueber die sensationellen Hintergründe

Ein ungewöhnlicher Schritt

Professor Fillunger hat diese Schrift, wie aus der Umgebung Professor Terzaghis erklärt wird, an eine Anzahl von Baufirmen in der ganzen Welt und an die internationalen technischen Institute versendet. Namentlich in Amerika, wo Professor Terzaghi im Juli des vergangenen Jahres als Präsident des ersten Kongresses für Erdbaumechanik fungierte, war diese Streitschrift in großer Auflage verbreitet worden.

Diese Broschüre Professor Fillungers erregte begreiflicherweise das größte Aufsehen. Sie löste

Report of the suicide of P. and M. Fillunger

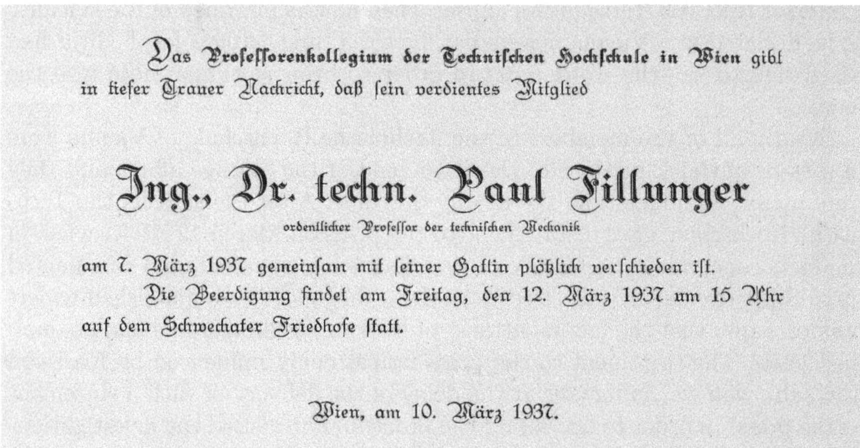

Das Professorenkollegium der Technischen Hochschule in Wien gibt in tiefer Trauer Nachricht, daß sein verdientes Mitglied

Ing., Dr. techn. Paul Fillunger

ordentlicher Professor der technischen Mechanik

am 7. März 1937 gemeinsam mit seiner Gattin plötzlich verschieden ist.

Die Beerdigung findet am Freitag, den 12. März 1937 um 15 Uhr auf dem Schwechater Friedhofe statt.

Wien, am 10. März 1937.

Obituary notice from Technische Hoschschule in Vienna

Professor Dr. Fillunger and his wife Margarete were buried in the Schwechater cemetery on March 12, 1937. At the grave, the Rector of the Technische Hochschule spoke the farewell words, in the name of the council of professors. The funeral was attended by delegations from the *Ingenieur- und Architekten-Verein*, by the council of professors, and by Fillunger's students.

The Technische Hochschule of Vienna received condolences from many universities and technical universities, e.g., of Vienna, Graz, Innsbruck, Salzburg, Aachen, Breslau, Brno, Danzig, Munich, Bonn, Frankfort on the Main, Freiburg, Halle-Wittenberg, Jena, Cologne, Leipzig, Prague, and Leoben.

The expert opinion paper was completed on April 9, 1937, and consisted of three sections: I. Befund, II. Gutachten, III. Begründung des Gutachtens (I. Findings, II. Opinion, III. Foundation of the opinion), also containing four appendixes. In the first section, the history of the affair was briefly sketched and the main points of the accusation were stated:

"Mr. Fillunger, in his polemic pamphlet, states that the theory of Mr. Terzaghi and Mr. Fröhlich is nonsense and denotes the laboratory tests as an impossibility. He introduces his own theoretical approach, which he cannot prove in practice."

In the second section, the expert opinion was formulated. At the very beginning of this section, it was stated that Fillunger was competent as a critic only in the theoretical field. Moreover, it reproached him for overlooking, in his treatise *Erdbaumechanik ?*, the theoretically important fact that, for the special dynamic problem which plays a decisive role for the time dependent settlement of clay layers, an initial value-problem had to be solved. The committee was of the opinion that this could only be performed generally with linear differential equations. Concerning the kind of representation in von Terzaghi's and Fröhlich's book, the committee came to the conclusion that Fillunger's criticisms were partly justified. However, the poor representation was an unimportant factor and did not play any important role in further investigation.

The ideas of von Terzaghi and Fröhlich were often misunderstood by Fillunger. This, one can suppose, was partially the reason for the factual mistakes and contradictions in Fillunger's brochure.

Even in the theoretical field, the committee stated that Mr. Fillunger was not right in view of the statements which referred to the subject. Even worse were his attacks against the laboratory tests and the applications in foundation engineering, which did not hold either.

The investigation committee remarked, furthermore, concerning the pamphlet *Erdbaumechanik ?*: "This treatise does not consider the usual fundamental rules for scientific publications which are characterized by the setting up of clear statements and by subsequent proofs. Moreover, the literature is only partially considered; also imperfections concerning the rendering of ideas and texts can be found." Although these points of criticism are also valid for the book by von Terzaghi and Fröhlich (see Appendix C of the expert opinion) and, in particular, for other papers written by von Terzaghi, the investigation committee did not mention this fact in the opinion.

In the third section on the foundation of the opinion, the committee discussed the first and second sections in Fillunger's brochure. It was recognized by the committee that, in the second section of Fillunger's treatise, new contributions were contained which could be used to treat the same problem in another way. However, with the assumption of quasi-stationary flow, one essentially obtained no other differential equation than that of von Terzaghi and Fröhlich (the calculation was contained in Appendix A). Appendix A was based on the calculations by Professor Flamm, who had used in his contribution important investigations by Professor Lechner and Dr. Heinrich, a co-worker of Professor Lechner. Flamm pointed out that those terms, in Fillunger's general system of field equations, which contain the porewater pressure had to be set right. However, Flamm's "corrections" consisted only of a simplification of Fillunger's general balance equations of momentum and mass towards the geometrically-linear theory.

The committee then considered the third section of Fillunger's pamphlet, which contained the integration of von Terzaghi's differential equation and some conclusions. The committee recognized that one could derive the integral formula without any neglect, in a shorter way than Fröhlich had done. Appendix B yielded the exact derivation. Moreover, Appendix B weakened all attacks from the third section of Fillunger's treatise. However, in Appendix B the committee criticized the kind of representation of the integration process contained in von Terzaghi and Fröhlich's book.

In the fourth section of his brochure *Erdbaumechanik ?*, Fillunger made the most severe errors. The committee showed this in detail in Appendix C of the opinion paper. In his pamphlet, Fillunger had given the impression via a self-invented and false formula which he denoted as the "Terzaghi-differential", that differential calculus was violated in the book by von Terzaghi and Fröhlich. However this was unfounded.

In Appendix C, the investigation committee completely adopted the arguments which Dr. Fröhlich had developed in the rebuttal, *Erdbaumechanik und Baupraxis: Eine Klarstellung.* The committee severely criticized, however, also in Appendix C, the representation of the theory in the book by von Terzaghi and Fröhlich (1936):

"It must, however, also be said that although the book 'Theorie der Setzung von Tonschichten', carries the subtitle 'An introduction to analytical clay mechanics', the reader becomes acquainted for the first time with the pore number through the following sentence on p. 24: 'By the porenumber ... one understands, as everybody knows, the fraction of the pore volume ... to the volume of the solid final matter ... of a given soil sample.' However, the definition is meant given in this Appendix. The aforementioned sentence is an unsuccessful formulation of this idea and easily confuses the reader."

Concerning Fillunger's criticism of the experimental investigations of von Terzaghi and Fröhlich, the committee refused all points of the attack and stated that the test methods of von Terzaghi's laboratory were correctly developed. Moreover, the time dependent settlement curve, based upon the relationship between the pore number and the grain-to-grain pressure, was in agreement with calculation and observation, as had been recognized during the committee members' visit to von Terzaghi's laboratory.

Section V of Fillunger's polemical pamphlet treated the construction technique of soil mechanics and, in particular, that of the theory of consolidation. The investigation committee stated: "Since Fillunger considers the theory as false, he also declares the application of the theoretical results in practice as impossible." The committee showed that, with the results of von Terzaghi's and Fröhlich's theory, an engineer was able to draw the right conclusions from drill logs in order to recognize instinctively the possibility of settlements, and to prevent later surprises. Moreover, the committee stated that this remarkable advance had not been considered by Mr. Fillunger. In Appendix D of the opinion paper, a list of research institutes from all over the world working in the field of soil mechanics was included, which was enough to reveal that Fillunger's criticism concerning construction techniques was unjustified.

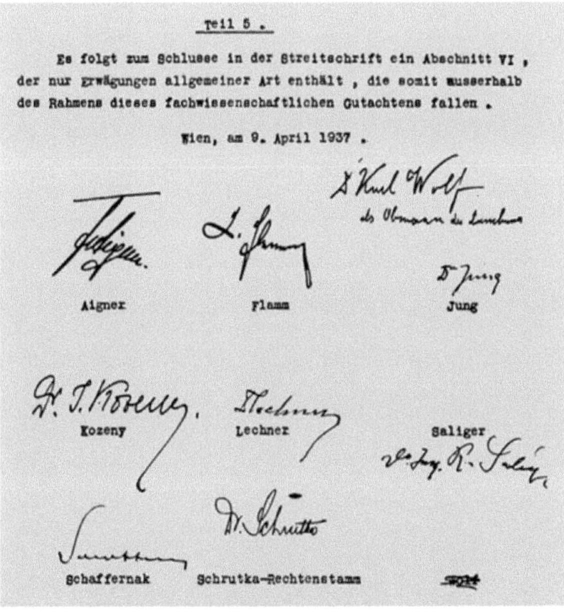

Teil 5 .

Es folgt zum Schlusse in der Streitschrift ein Abschnitt VI , der nur Erwägungen allgemeiner Art enthält , die somit ausserhalb des Rahmens dieses fachwissenschaftlichen Gutachtens fallen .

Wien, am 9. April 1937 .

Aigner Flamm Jung

Kozeny Lechner Saliger

Schaffernak Schrutka-Rechtenstamm

The expert opinion signed by the members of the investigation committee

The final section of Fillunger's brochure was not discussed by the committee because of the fact that it contained only general reflections which lay outside the framework of the expert opinion.

The opinion was signed by the members of the investigation committee on April 9, and subsequently put under seal. Only individuals having an authorized interest in the entire case were permitted to read the opinion.

With the suicide of Paul Fillunger, the signed experts opinion, and the formal end of the disciplinary action during the summer of 1937, the fight with Karl von Terzaghi was essentially decided. How far this conflict went on to influence the further development of the porous media theory will be discussed in the next sections.

4.2 The Aftermath

In July, 1937, the revised version of *Vorläufige Erwiderung auf die Verteidigungsschrift der Herren v. Terzaghi und Fröhlich* by Paul Fillunger was edited by his son Erwin Fillunger, with the new title, *Erdbaumechanik und Wissenschaft – eine Erwiderung* (*Soil mechanics and science – a reply*). In the preface Erwin Fillunger pointed out: "The paper was in the existing form when my good father unfortunately came to the terrible conviction that he was wrong. Due to his immediate inexorable decision, the printing remained unfinished.

I am editing the paper now in the responsibility to fulfil with this a serious filial duty of thankfulness towards the memory of my beloved parents. The publication takes place in the smallest edition and only for the purpose of making it possible for the friends of my father to also become acquainted with his last standpoint."

This last paper of Fillunger did not reach the public and has not been cited in the literature.

At the very beginning of January, 1938, Dr. Heinrich (1938), later also professor at the Technische Hochschule of Vienna, published a paper entitled *Wissenschaftliche Grundlagen der Setzung von Tonschichten* (Scientific fundamentals of the theory of the settlement of clay layers), in which he presented the findings of Professor Lechner and himself, which had been discovered within the framework of their co-operation in the investigation committee in the first months of 1937. Heinrich (1938) pointed out in the introduction:

"The differential equation developed by Professor v. Terzaghi, which is the basis for the theory of the settlement of clay layers, carries consciously the character of a first approach to the problem raised. The approach may be sufficient in many cases for the calculation of practical soil mechanics. Due to the fundamental importance of a differential equation describing the settlement process; the demand for a more exact treatment remains, however, open.

In the following it will be attempted to give the fundamentals for a stricter theory of the settlement, whereby in particular a high value shall be set on bookkeeping exactly over the introduced assumptions if it is possible. It shall also aim to make the necessary approximations available to a numerical estimation.

Professor Dr. P. Fillunger has already pointed out in his paper *Erdbaumechanik?* that the settlement in connection with the porewater flow is a matter of a coupled motion of two phases, a solid and a liquid. This assumes, however, that the pore space between the solid substance of the clay is totally filled with water. The treatment chosen here follows the representation in the paper mentioned, regarding the fundamental form of the advanced equations."

ERDBAUMECHANIK
UND WISSENSCHAFT

EINE ERWIDERUNG

VON

BAURAT PROF. DR. PAUL FILLUNGER
WIEN

ALS MANUSKRIPT GEDRUCKT

DRUCK UND VERLAG DER
BUCHDRUCKEREI FRIEDRICH JASPER, WIEN, III., THONGASSE 12

P. Fillunger's reply to the defense pamphlet of Karl von Terzaghi and O.K. Fröhlich, edited by his son

Moreover, in the conclusion to his paper, Heinrich (1938) stated: "It is attempted to realize a treatment of the problem of the settlement of clay layers on a scientific basis, ..."

This statement corresponds to Fillunger's question (see Fillunger, 1937) in his response to the defending paper by von Terzaghi and Fröhlich (1937), namely whether von Terzaghi's differential equation could be scientifically founded.

Professor Flamm was very indignant about Heinrich's comments. Flamm had already visited von Terzaghi on January 24, 1938, in order to discuss Heinrich's paper. In the next days, Flamm (1938) wrote a treatise which he called *Beitrag zur Theorie der Setzung von Tonschichten* (Contribution to the theory of the settlement of clay layers). The manuscript appeared in May, 1938, in the same journal as Heinrich's paper. In the introduction, Flamm (1938) cited Heinrich's aforementioned statements and commented angrily with: "For the reader this could easily give the impression that the calculations by v. Terzaghi would not be scientifically correct. The following lines shall determine that such an opinion is unfounded, and will clarify also the other statements from Mr. Dr.-Ing. Heinrich."

It is obvious that Flamm's reply to Heinrich's paper was decisively influenced by von Terzaghi. There are several statements in Flamm's paper which "smell" of Karl von Terzaghi's influence.

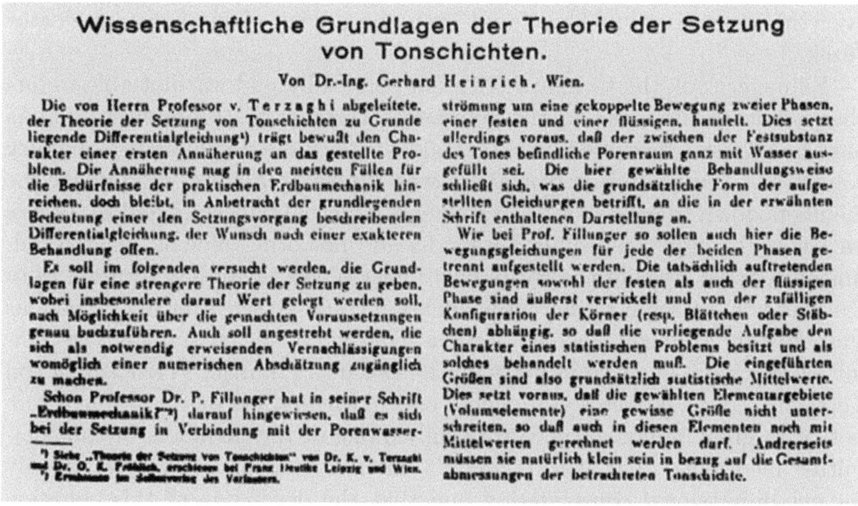

Wissenschaftliche Grundlagen der Theorie der Setzung von Tonschichten.

Von Dr.-Ing. Gerhard Heinrich, Wien.

Die von Herrn Professor v. Terzaghi abgeleitete, der Theorie der Setzung von Tonschichten zu Grunde liegende Differentialgleichung[1]) trägt bewußt den Charakter einer ersten Annäherung an das gestellte Problem. Die Annäherung mag in den meisten Fällen für die Bedürfnisse der praktischen Erdbaumechanik hinreichen, doch bleibt, in Anbetracht der grundlegenden Bedeutung einer den Setzungsvorgang beschreibenden Differentialgleichung, der Wunsch nach einer exakteren Behandlung offen.

Es soll im folgenden versucht werden, die Grundlagen für eine strengere Theorie der Setzung zu geben, wobei insbesondere darauf Wert gelegt werden soll, nach Möglichkeit über die gemachten Voraussetzungen genau buchzuführen. Auch soll angestrebt werden, die sich als notwendig erweisenden Vernachlässigungen womöglich einer numerischen Abschätzung zugänglich zu machen.

Schon Professor Dr. P. Fillunger hat in seiner Schrift „Erdbaumechanik?"[2]) darauf hingewiesen, daß es sich bei der Setzung in Verbindung mit der Porenwasserströmung um eine gekoppelte Bewegung zweier Phasen, einer festen und einer flüssigen, handelt. Dies setzt allerdings voraus, daß der zwischen der Festsubstanz des Tones befindliche Porenraum ganz mit Wasser ausgefüllt sei. Die hier gewählte Behandlungsweise schließt sich, was die grundsätzliche Form der aufgestellten Gleichungen betrifft, an die in der erwähnten Schrift enthaltenen Darstellung an.

Wie bei Prof. Fillunger so sollen auch hier die Bewegungsgleichungen für jede der beiden Phasen getrennt aufgestellt werden. Die tatsächlich auftretenden Bewegungen sowohl der festen als auch der flüssigen Phase sind äußerst verwickelt und von der zufälligen Konfiguration der Körner (resp. Blättchen oder Stäbchen) abhängig, so daß die vorliegende Aufgabe den Charakter eines statistischen Problems besitzt und als solches behandelt werden muß. Die eingeführten Größen sind also grundsätzlich statistische Mittelwerte. Dies setzt voraus, daß die gewählten Elementargebiete (Volumselemente) eine gewisse Größe nicht unterschreiten, so daß auch in diesen Elementen noch mit Mittelwerten gerechnet werden darf. Andrerseits müssen sie natürlich klein sein in bezug auf die Gesamtabmessungen der betrachteten Tonschichte.

[1]) Siehe „Theorie der Setzung von Tonschichten" von Dr. K. v. Terzaghi und Dr. O. K. Fröhlich, erschienen bei Franz Deuticke Leipzig und Wien.
[2]) Erschienen im Selbstverlag des Verfassers.

Heinrich's first paper on the porous media theory

With Flamm's (1938) paper, the scientific conflict between Fillunger and von Terzaghi was finished.

The reader may put the question as to whether or not Fillunger had a chance to overcome the disciplinary action as an innocent person and to win the fight against von Terzaghi. The answer is an absolute "no."

Fillunger lived in a very small world of his own; his whole life – schooling, studies, as well as his occupational career – was confined mainly to the city of Vienna. He was very deeply involved in his scientific research, which he very much overrated, and had a certain arrogance; in a nutshell, he was a good scientist however, a know-it-all who was always busy teaching others, and all too often did not accept any other opinion.

In contrast to Fillunger, von Terzaghi was an internationally known personality, a cosmopolitan, having made international trips to many countries. Science was not the only field which occupied him; he was very busy as a consultant, which in turn gave him the opportunity to interact with more people. He was a very perceptive person, having the ability to sharply analyze problems as well as to recognize the important issues. Furthermore, he was sociable, charming, and easy-going in his dealings with people, which, in turn, gave him an attractive personality.

Von Terzaghi was much better equipped for the controversy compared to Fillunger. He used his aforementioned capabilities to build up his defense in which he won over the "right" people, who later promoted his ideas in the committee, and he took the advice of experts, in particular, in legal problems, even of Fritz Byloff his brother-in-law, whereas Fillunger excercised the so-called *Caesarian mania* of the professors at that time, namely to do everything on his own; he refused even the advice of his assistant doctor Jêzek.

Fillunger took the technical criticism personally and was not able to further advocate his ideas to the committee through his representatives. As has already been discussed, von Terzaghi's supporters dominated the inquiry committee; the leading person there was Professor Flamm. He pinned the members down to concentrate on Fillunger's factual offensives and to discuss only von Terzaghi's consolidation model and not other models, such as Fillunger's. At no stage of the entire investigation was the question raised as to whether von Terzaghi's differential equation could be scientifically proven, as suggested by Fillunger. Instead of concentrating on this main point, the committee was more involved in petty matters such as Fillunger's criticism of von Terzaghi's and Fröhlich's statements and mathematical details, which were not of as great importance. Surprisingly, none of the members of the committee mentioned that von Terzaghi's differential equation was valid only for one-dimensional consolidation and that the derivation of this equation was inadequate. Likewise, it is also very surprising that no one perceived (or wanted to perceive) that Fillunger had founded a general mechanical theory for saturated porous media. It appears, in some respect, perfidious that Fillunger's basic ideas and his field equations served to show that von Terzaghi's differential equation was equal to Fillunger's field equations. In the expert opinion, no hint appeared that von Terzaghi had derived his differential equation by an analogous conclusion and not from the fundamental balance equations of mechanics. Moreover, in order to reveal the "equality" of both approaches, Flamm manipulated Fillunger's field equations by setting the volume fractions constant, which is only approximately valid within the geometrically linear theory. He maintained, in the Appendix A of the expert opinion: "The terms containing the porewater pressure ... must, however, be set right; the equations have to read": and the manipulated equations of motion followed after this remark. With this scandalous manipulation, Flamm obviously pursued two purposes: first, only with his manipulated equations could he derive von Terzaghi's differential equation; second, with his manipulation he could bring Fillunger into disrepute. It is not known how far Flamm's action was arranged with von Terzaghi.

Clearly, it seems as if the representatives from the field of mechanics in the inquiry committee had not paid attention to Flamm's procedure and let themselves be misled by Flamm. It seems as if they had a guilty conscience, otherwise the prompt publication of Heinrich's (1938) paper at the very beginning of 1938, in which he criticized, in a somewhat hidden form, von Terzaghi's scientific working, cannot be explained. In his paper, Heinrich simplified, in a very correct form, Fillunger's field equations to the geometrically-linear case, thus arriving at the manipulated formulas of Flamm. Flamm (1938) wrote, in his rebuttal, that Heinrich was wrong on many points and that Heinrich's merit would consist of setting right Fillunger's balance equation (which had not been done by Heinrich). Thus, he used the same words as in the Appendix A of the expert opinion in 1937; from this, we can conclude that Flamm's manipulation of Fillunger's equations does not seem to be a slip. This can be stressed by the point that Flamm visited von Terzaghi at least two times to discuss his reply to Heinrich's paper.

It seems that not only the representatives from the field of mechanics had a guilty conscience but also Professor Flamm; he never spoke about the sad affair, even to his close family members and later, in 1950, he strongly and successfully supported Heinrich's application for the vacant professorship (of professor Wolf) for mechanics at the Technische Hochschule of Vienna.

Fillunger's confidant on the inquiry committee, Professor Saliger, described the affair in his autobiography under the title *The idealist Fillunger against the practical man Terzaghi.* He remarked therein: " Without doubt Fillunger's disclosure about certain events in the scientific world cannot be generalized. How justified however its core is in single cases, is shown by the following example" and he reported on a similar case in his own field of concrete construction.

After the bizarre affair, there was a rumor, according to the relatives of Fillunger, that von Terzaghi was requested by Fillunger to stop the disciplinary action. Von Terzaghi, however, completely rejected Fillunger's request.

Karl von Terzaghi, 1937

How the strange Fillunger-von-Terzaghi affair was seen by the late Ruth D. Terzaghi, the second wife of Karl von Terzaghi, can be read from a statement in a letter to the author from December 13, 1989: "I believe that Arthur Casagrande gave the matter (the Fillunger-von-Terzaghi-affair, the author) such slight attention because no one in this country took the matter seriously. Fillunger was regarded as, at best, a freak and, at worst, as a psychotic."

As has already been pointed out, Fillunger's brilliant findings were nearly totally ignored in the literature, a sad fact which was partially enforced by the disciples of von Terzaghi. For example, within the framework of the edition of the book, *From theory to practice in soil mechanics* (Bjerrum, Casagrande, Peck, and Skempton, 1960), Professor Skempton sent a first draft of his article *Significance of Terzaghi's concept of effective stress* to Professor Arthur Casagrande, where he mentioned Fillunger's contribution, referring to Fillunger's paper from 1915 on the discovery of effective stresses, with the words:

"... yet neither he nor anyone else at the time realized the full significance of these results."

A. Casagrande answered Skempton very angrily in a letter of June 9, 1959: "I am still not happy about the manner in which you bring Fillunger into the picture. I don't know whether you have ever read Fillunger's booklet which he published for the purpose of discrediting all that Terzaghi has done. Not only was he scientifically completely wrong but it was such a vicious and uncalled for attack in which he ridiculed the principles of soil mechanics that were established by Terzaghi. You are still giving him credit where no credit is due, by stating: 'This amounts to a corollary of the principle of effective stress', even though you go on to say that 'neither he nor anyone else at the time realized the full significance of these results'. A reader may gain the impression that in a vague way Fillunger had understood the mechanics but had not yet realized the full significance. The fact is that he denounced as nonsense Terzaghi's concept of pore pressures and effective stresses. The least I would ask you to do is to delete the word 'full' in the sentence '.. yet neither he nor anyone else at the time realized the full significance of these results'."

In a letter to A. Casagrande of June 22, 1959, Skempton agreed: "Regarding your comments on my revisions, I will certainly omit the word 'full' in the sentence to which you refer ..." Indeed, in the final form of the book in question, the word "full" was omitted.

The Fillunger-von-Terzaghi-affair greatly influenced the further development of porous media theories and, in particular, the development of soil mechanics, in a very negative sense.

Fillunger's work was, as aforementioned completely forgotten, ignored and even consciously passed over in silence due to the affair described above, although his theory was well ahead of its time. It is a pity that his followers Heinrich and Desoyer failed to reach the international scientific community with their papers of considerable value, partly – without doubt – owing to the fact that they published their results in German.

The whole affair with Fillunger also had a great influence on Karl von Terzaghi's further work.

Already in the early days of his academic carrier, von Terzaghi had had a skeptical view of any kind of theory. However, on the other hand he knew that he could not establish a new field in engineering sciences without doing theoretical work. However, he was limited in his ability to work scientifically because of his restless lifestyle, his sometimes unconcentrated, flighty working on theoretical topics and his lack of basic mechanical and mathematical knowledge as well as his disinterest in bringing his physical discoveries into the right written form, comprehensible for the reader – and being duplicated.

It seems that the whole affair with Fillunger had shocked Karl von Terzaghi, who originally wanted to become, among other things, a mechanics professor, probably in order to follow in Professor Wittenbauer's footsteps. After 1937 he stopped his theoretical research completely and turned exclusively to practical problems and to the observational method. He justified this in 1948 (see two paragraphs further, the author) when he declared that basic research in soil mechanics had already come to an end in 1936. However, this conclusion does not correspond to a statement in 1935 when he asked von Kármán, who was involved in the preparation of the great mechanics conference of the International Union of Theoretical and Applied Mechanics (1936) in Cambridge (USA), not to include soil mechanics as a special session in the conference because "it was not ripe."

It may be that the confrontation with attacks from a theorist had convinced him to completely stop scientific work in the theoretical field. In any case, after leaving Vienna in 1938, he promoted only one student at Harvard University, where he had taken over a visiting lectureship. From that point on, no substantial contribution by von Terzaghi to the theory of the multicomponent soil body is known. All of his further contributions to soil mechanics were devoted to the practical, engineering-scientific side of soil mechanics. He was also of the opinion that laboratory work and the theoretical aspects of soil mechanics had already been brought to a close. In the opening address at the Second International Conference on Soil Mechanics and Foundation Engineering (Rotterdam) in 1948, von Terzaghi pointed out:

"Every science, pure or applied, is based on what is known as fundamental research. Prior to 1936 fundamental research in soil mechanics consisted chiefly in the investigation of the significant soil properties by laboratory tests, and in the development of theories of earth pressure, stability and settlement. In 1936, this pioneering stage of soil mechanics research was already completed. It had created what may be called an ideal pattern of soil behavior and it had placed at the disposal of the practicing engineer a set of theoretical concepts covering every important field of soil behavior. The concepts were based on the laws of applied mechanics and on the results of laboratory tests performed under rigidly controlled conditions on soil samples which were believed to be almost undisturbed."

Moreover, in the outlook of his opening address, he stressed his standpoint:

"The days in which significant discoveries could be made in the laboratory, or at the writing desk, appear to be gone forever."

His view, and his criticism of the theoretical penetration of soil mechanics, were summarized in a somewhat ironic tone in a letter to Mr. Harding (England) in 1952:

"I greatly enjoyed your occasional reference to the mathematical mind. I would define a mathematician as a man who can solve every equation but he may fail to notice it when he displaces a decimal point. Once I assigned to a mathematically minded engineer, of high academic standing, the following task. He should estimate the flow of water into an open excavation for the different stages of excavation to bedrock at a depth of about 100 ft. He promptly set up the differential equations which were rather involved and one week later, full of pride, he presented the results. According to the results of his computations – which were correct provided you endorsed the assumptions – the inflow is a maximum at a depth of about 60 ft. As excavation proceeds beyond this depth the inflow decreases and becomes very moderate at the final stage. 'Fine', I said. 'Now you work out a procedure for excavating down to 100 ft without passing through that confounded intermediate stage in the proximity of 60 ft'. Grand fellows: I have suggested repeatedly that we should keep them in cages and feed them their problems through the bars. If there is no danger that they may get out they can be quite useful.

Once every few months, I get a manuscript full of differential equations describing a revolutionary discovery in the realm of theoretical soil mechanics, for review and comments. I have composed a sort of standard letter for my reply and the manuscripts are tucked away in a fat file labeled 'Nut House'."

It seems that Karl von Terzaghi – after the supposed great victory over his opponent Paul Fillunger – had lost any doubt about the correctness of his scientific work. His self-confidence had grown out of all proportion. How meager, however, his tour of mountain ridges in the scientific field was, can be revealed evaluating von Terzaghi's scientific achievements. Only his "physical scent" – as he called it – hindered him from slipping from the ridge in several cases. Without doubt his performance in correcting Fillunger's uplift formula and approximately formulating the exact value was a success owing to his ability to recognize physical effects. However, the proofs of 1934 and 1937 are not founded on any physical theorems and must be rejected. The proof shows that he had not really understood his result, because already in 1922 he had published the correct uplift formula for sand, however, without any proof or reference to other papers.

5. The Further Development

5.1 Karl von Terzaghi's Life and Work after the Sad Affair

After the sad affair with Paul Fillunger, Karl von Terzaghi began again to look into the uplift problem. From March 10 to March 20, 1937, he worked relentlessly on this problem, according to the Oslo version of his diary. Then, on March 22 he stopped his work and stated in his diary: "Stopped uplift. Recognized that μ_{max} [53] for dams inadmissible." This reason for stopping his research on the uplift problem is not convincing.

Fortunately, von Terzaghi prepared a manuscript, which was never published. This manuscript was found in the Terzaghi Library at the Norwegian

[53] Effective surface porosity.

Karl von Terzaghi, 1937

Von Terzaghi's apartment in Vienna, Kahlenbergstraße 59.

Geotechical Institute (NGI) in Oslo. It has the title *Die mechanische Wirkung des Auftriebes in porösen Körpern* (The mechanical effect of uplift in porous bodies).

In the paper in question the uplift problem is discussed extensively and is accompanied by some considerations on the strength and the deformations of a porous column. He repeated his criticism of Fillunger's theory and stated the approximately correct form of the uplift formula. The strange derivation and proof of his formula must, however, be rejected, because they are not based on fundamental mechanical principles. He was, namely, of the opinion that uplift depended on the contact area between the grains. If the contact area is negligible compared with the total surface of the grains the full hydrostatic uplift is acting, otherwise, the uplift is reduced. Karl von Terzaghi's statement contradicts clearly Archimedes' principle. He tried to prove his statement by experiments and strength hypotheses. The result was that he obtained approximately the same result, namely the result by using Archimedes' principle. In his book *Theoretical Soil Mechanics* from 1943 he disclaimed all proofs.

In April, 1937, Karl von Terzaghi worked hard on his lecture on the *Reichsbrücke* which he presented at *The Österreichische Ingenieur- und Architekten-Verein* on April 30. The reception was very cool. Obviously, the audience had not forgotten the sad Fillunger-affair. However, the talk broke the ice. On Monday, May 10, Karl reapeated the talk in Graz, where only few listeners were present: three or four from the Technische Hochschule, some acquaintances – and his family.

From Graz, Karl and Ruth started off for a visit to Croatia, to the Dalmatian coast, in order to relax and to recover from the strenuous Fillunger affair. They spent restful days in Agram (Zagreb), Split, some small towns on the coast and returned to Vienna at the end of May. After the vacation, Karl von Terzaghi travelled to the central part of Sweden in order to deliver an opinion paper on a conflict between the city of Stockholm and the Swedish Goverment. In the middle of June, he returned to Vienna and the von Terzaghi family moved from the *Dollfußplatz* to their small new apartment in Grinzing in the 19. District, Kahlenbergstr. 59. "A large room, separated by a curtain in two parts, a living room and a bedroom. On the north side a small glassed-in porch as office, on the south side a small bathroom and a children's room. Which was filled up comfortably by Ruth in a few days."

His work at the Technische Hochschule was pleasantly interrupted several times by visitors from abroad.

Also the Academy of Sciences closed the book on the Fillunger affair. The members elected Karl von Terzaghi as a real member of the Academy in June, 1937.

From July 10 to 18, Karl attended a student excursion through the southern part of Germany organized by Professor Saliger: Berchtesgaden, Munich, Augsburg, Ulm, Heidelberg, Mannheim, Würzburg, Bamberg and Nuremberg, where they visited many construction sites in the fields of road and water construction. The excursion, generously supported by the Generalinspektor Dr. Todt, was a success, on account of Saliger's excellent planning.

Following the excursion, von Terzaghi's consulting activities grew steadily. He examined several landslides at the Aachensee, became a consultant of the French firm *Les Travaux Souterrains*, and gave his opinion on the foundation of a power plant in Latvia in September, 1937.

In October, 1937, von Terzaghi travelled with his wife to Paris in order to hold a speech there and to give advice to *Les Travaux Souterrains* concerning the stability of a barrage at Béni Bahdel in Algeria. At the end of October, the von Terzaghis returned to Vienna. In the two months to follow, he had enough leisure time to work out his lecture notes on soil mechanics. However, by the end of December, he was on his way to London in order to do some consulting work and, in January, 1938, he was studying the first results concerning the observations of the barrage at Ghrib, a city near Algier. After his return to Vienna, von Terzaghi wrote up the final results and soon was able to deliver his opinion paper to the directorate of *Travaux Souterrains*.

On March 3, 1938, von Terzaghi took over the honorable function of outlining the merits of the former President of the United States of America, Herbert Hoover, in economics and engineering in a ceremonial act at the Technische Hochschule of Vienna, and of conferring the honorary doctorate

K. von Terzaghi on an excursion through the southern part of Germany (Schaffernak fourth, von Terzaghi eighth from the left)

degree on him. On this occasion, the high officials of Schuschnigg-Austria gathered for the last time and the festival hall was completely full.

In March, 1938, the political situation in Austria changed radically. Post-world-war Austria ceased to exist. After the collapse of the Habsburg monarchy in 1919, Austria had to renounce claims, under pressure from the victorious nations, to the Southtyrol and Sudetic Mountains and had to recognize the new states of Czechoslovakia, Poland, Hungary and Yugoslavia. Moreover, besides the payment of high reparations, Austria had to declare the renunciation of a union with the German Reich. For many Austrians this renunciation was the hardest condition imposed by the victorious nations. After the great territorial loss, the old longing of many Austrians for the unification with Germany came to life again. At the end of the 1920s a plan to set up a customs union with Germany failed, the unemployment rate grew dramatically, crises in the Parliament and insufficient cooperation between the parties led to calls for antidemocratic and radical solutions, in particular, within the circles of the young and middle-aged generation. Many citizens joined the front-line-union, the militia and the NSDAP[54]. On the left the half-military Republican Defense Alliance was created. Gradually the do-

[54] Nationalsozialistische Deutsche Arbeiterpartei (National Socialistic German Workers Party) (better known as the Nazi Party).

Honorary Doctor degree
for President H. Hoover.

mestic political situation got worse. Dollfuß, who was ordered to form the new government in 1932, tried to save the situation through reforms of the Parliament and the constitution. He used the resignation of the three Parliament presidents in March, 1933, to eliminate the Parliament. He prohibited the activities of the NSDAP and disbanded the SA and SS[55]. In February, 1934, the Republican Defense alliance, also disbanded, started a revolt which was put down brutally by the militia and military. On May 1, 1934, a feudal constitution was proclaimed. Meanwhile Chancelor Dollfuß was so hated by the NSDAP that members of the party assassinated him during a putsch attempt. The new Chancelor, Schuschnigg, tried to continue Dollfuß's policies. However, the Nazis had widened their terror and had infiltrated the official machinery and had thus manoeuvred Schuschnigg into a helpless position. Berlin used this situation and Hitler invited Schuschnigg to come to Berchtesgaden, to the Obersalzberg to discuss the situation in Austria. The meeting took place on February 12, 1938. Hitler treated Schuschnigg very rudely and delivered an ultimatum to him which contained, among other things, the following demands: the prohibition of the Austrian NSDAP must be cancelled; all arrested supporters of Hitler must be pardoned; and Dr. Seyß-Inquart, a Viennese lawyer closely related to the National Socialists must be appointed

[55] Sturmabteilung and Schutzstaffel, security arms of the Nazi Party.

Minister of the Interior and, thus, would be responsible for the police and security.

On March 11, Schuschnigg cancelled the intended referendum under pressure from Hitler and resigned. He left his function to Dr. Seyß-Inquart. On March 12, 1938, German troops entered Austria and on March 13, the annexation was carried out.

Karl von Terzaghi devoted several pages in his diary to the great event from his point of view. "Friday, March 11. During last days feverish preparation for 'Volksabstimmung'(plebiscite), poor VF[56] demonstrations in the streets in the evening ... It is rumored that Schuschnigg will mobilize Schutzbund (defensive alliance) on election day. On Friday morning I found at the Technische Hochschule a circular letter that the professors have to vote in public at the Technische Hochschule on March 12. That was too much. Arranged with Haas[57] and Schaffernack. Should bring together a small group, which refuses the vote. Estimated maximum of a dozen people among, them Hartmann[58]. Meeting of the ringleaders in a coffee house at noon. In the afternoon it was adviced that we should wait for the evening. Seyß-Inquart would give instructions in the Deutschen Klub (German Club). Discussed the facts with the assistants. They should do what they think is best. I will not vote.

In the evening, at home in the Kalenbergstraße. Around 10 p.m. telephone call, Mitzi (Marie Obermayer, jurist in the Zentral-Europaische Länderbank, Vienna) with trembling voice: Schuschnigg has resigned. He is supposed to have said via radio: 'I give way to the violence and take my leave of the Austrian people! German troops have crossed the border.' Ruth jumped for joy. The dissembling Schuschnigg system is gone. We drank a schnaps in honor of the hour and invited the fat female cook, Mitzi (Marie Steinlechner), Tyrolese and arch-Nazi, to participate.

Saturday, March 12. The Technische Hochschule looked as if transformed. In place of police officers, students in SA uniforms at the entrance. Swastika flags everywhere, even in my office. My assistants in the third floor: Kottnigg (?) in SA uniform, Brausewetter as a slender SS man in black. Everybody already assigned to his duty. As if one had already been drilled for weeks. Waited around noon to no purpose in front of Schaffernack's radio for Hitler's speech in Linz. His arrival was shifted from one hour to the next. In the turbulence Fröhlich came in; he was on his way from Mannheim to Istanbul. Travelled with night train and learnt first in Vienna what had happened. Supposed to rush on in the evening. Borders closed – little chance to get through Hungary. Indeed all Jews were carted away. He got through.

Monday, March 14, 9. a.m. in the Technische Hochschule. Professors meeting appointed for 9:30 a.m. by mistake. Professors gathered in the room be-

[56] Vaterländische Front (National Front).
[57] Professor Haas was an architect and Nazi.
[58] Professor Hartmann was professor for steel construction.

Austrian people welcome German troops

hind the tables, Rector Holey came from the rectorate office and tried to correct the error. Lösel, the small spitfire, jumped out of the crowed, chewed out the rector, he has nothing to say here anymore, as a Black[59]. New Era has started! Haas could calm him only with effort. When we left, Lösel dragged half a dozen colleagues into his office, opened a real Soviet. 'The lax attitude of the Hochschule during the *Machtergreifung* (assumption of power) in Germany must not happen again. He himself will take care for that. We listened to his barking silently and oddly for a while and left. In the meantime, Haas and Schaffernack negotiated with Holey in order to suggest to him to lay down his position as Rector in the meeting at noon. The Prorector Böck did the same and the old hypocrite Saliger, who had already prepared a well-worded ceremonial address, took control.

After the meeting I went to Holey, stirred up inside, in order to express to him my shame and anger about the Lösel incident, without success. We both cried. It was the hour of Austria's downfall.

There were also other members of the Technische Hochschule who were affected by the annexion, e.g. the Jewish Administrationsrat Dr. Josef Goldberg was immediately dismissed.

In the evening, torchlight procession at the *Ring*. Participation of the professors obligatory. Ruth was also with me. On the Dollfuß-Place a closely

[59] "Black" (Schwarzer) is an expression for a very conservative person.

People of Vienna listening to
Hitler's speech at the Heldenplatz.

packed crowd. The torchlight procession did not get on and we vanished already at the University. In the tram all the passengers were wearing swastika signs.

Tuesday, March 15, morning. The professors marched compactly to the Heldenplatz. There, students in SA uniforms served as ushers and led us across through all barriers with splendid discipline. The whole, wide Heldenplatz occupied by crowd of people, head to head, body to body. They hung like grapes on the monuments of the imperial generals. From time to time speaking choirs: One people, one empire, one Führer. –Who has saved us? Adolf Hitler! Hartmann, the bridge builder said, now he knows as to how he has to begin his lectures. With a speaking choir: "Now we will build a bridge ..." After one hour of waiting Hitler appeared, in the midst of his brown uniformed trabants, on the balcony of the new castle, decorated in red. The cheering of the people was so intense that Hitler had to wait for some minutes before he started his speech. His clear, far resounding organ trembling with excitement had a fascinating effect on the crowd. The people stood as if paralyzed and saw with the speaker the huge change in the fate of the Ostmark.

In the afternoon parade of the German troops. Also this time the streets were blocked by the crowd. We could approach the procession only up to 200 meters. The Technische Hochschule was like a camp. Each student and each assistant appeared to have his post. Haas appointed as liaison between party and Hochschule. Schober, a simple and unimportant Privatdozent is leader of all lecturers including the professors."

Despite all the joy about the annexation in wide circles of the people the bad and terrible sides of the Nazis became soon apparent. Karl von Terzaghi reported:

"Friday 18. with Dr. M., received me with his party badge. He has experienced already bad excess of the new system. A student visited the physicist Prof. E. (cosmic rays) and extorted 500 Schilings from him with the argument, E. has used his thoughts in a paper. People are arrested by SA men without any foundation and M. must make a great effort to get them released again. Gave me the advice to invest Ruth's means in such a way that it is not notifiable according to the German foreign exchange laws. To this belong real estate and silent partnership."

On Sunday, during a walk Mitzi told of the tragedies which are happening at the international bank. "Jewish officials are dismissed immediately. Again and again Gestapo officers appear and take away officials or clients to unknown places. The suicide epidemic among the victims has already started. In her house, second floor, in former rooms of the VF, the SA is accommodated.

On Jewish shops inscription *Jude*. Whoever buys there will be insulted or will be led through the streets with the inscription around his neck: this Arian swine has bought from Jews. Dr. P. has become completely down and apathetic. He is helpless."

At the end of March, Karl von Terzaghi sent his wife to Paris in order to invest her money either in a partnership or in real estate, leaving baby Eric in Vienna; care was taken by the nurse maid Erna. If Ruth failed in Paris, she was to take the next steamer to Boston.

By March, 1938, von Terzaghi was asked again to visit the barrage in Algeria, travelling this time via Paris. He struggled with the authorities to get a visa for France. Finally, he wrote to the Rector of the Technische Hochschule that he would need a visa for France. The temporary Rector forwarded von Terzaghi's request by letter to the police headquarters of Vienna on March 28, 1938. Obviously, von Terzaghi must have received his visa immediately, for he was on his way to Paris on March 29, arriving there on March 30. In Paris he was very busy in doing consulting work. Obviously, Ruth was not successful in Paris in investing her money, because she travelled further to Boston. Karl and Ruth exchanged several telegrams in order to clarify the permanent investing. In the beginning of April, he went via Marseille to Ghrib. The next days were filled with studies and discussions over the foundation of the barrage. In the middle of April, he returned to Paris where he worked hard during the following two weeks to complete his opinion paper on the barrage in North Africa. However, he also took time to enjoy Parisian life (the Casino de Paris). His stay in Paris was interrupted only by a visit to London (April 25 to 27) in order to give some advice in connection with difficulties encountered in the construction of an earth dam in England. On the evening of April 27 he was back in Paris and met his wife, after her return from Boston, in the hotel d'Antin, where he discussed with Ruth their plans for the near future, in particular, the emigration of Eric to Paris.

He was back at the Technische Hochschule on April 29, 1938. In his absence, he had missed a great part of the troubles and the ceremonies in Vienna in connection with Austria's annexation. In Vienna, von Terzaghi again took up lecturing in front of a few students on May 2 whereby he reported of his meeting with Hitler. During the days to follow he worked on the opinion paper for the barrage at Ghrib.

Obviously, Ruth did not want to return to Vienna. Thus, Karl had to decide what he should do. Baby Eric was still in Vienna and Karl would have to resign from a secure position with pension claim at the Technische Hochschule if moving to Paris. However, it seems that Karl came to terms with sending little Eric to Paris and that he himself would give up his position in Vienna. He discussed his problems with his colleague Schaffernak. However, Schaffernak only stated: "Today there is only the conclusion one hundred percent for it or against it." Another subject was much more interesting for Schaffernak. He had the plan to put von Terzaghi's and his institute in the service of the 'Bewegung' (Nazi movement). However the plan failed. In April Karl "gave notice with heavy heart giving up the flat in the Kahlenbergstreet." However he was not sure what to do. Therefore, he followed the advice of an acquaintance to visit the counsellor Etscheit in order to get some information concerning his situation. Unfortunately, Etscheit had left Vienna for Berlin. In the evening of May 9, Karl rode to Berlin where Rendulic was expecting him. On the next afternoon Karl von Terzaghi went to Etscheit's old fashioned office in the elegant boulevard *Unter den Linden.* Karl told him that he would prefer to have double residence, half a year in Paris or London[60], and half a year in Berlin with Todt's organizations and he asked the counsellor whether he could dispose of his means in foreign countries. However, Etscheit answered him that it would be extremely difficult to take up the double residence. In contrast the disposition of his means would be easier. In the time to follow Karl had some additional meetings with Etscheit in Vienna. The counsellor was very pleased with the first draft of a letter Karl had prepared for Dr. Todt requesting support for his efforts and to protect the interest of his wife. However, finally the counsellor could not raise hope concerning Karl's situation and Karl was forced to give up his flat.

Moreover, the atmosphere at home was severely disturbed. Karl discovered that Erna "the excellent and mature nurse" was harassed by Mitzi (Marie Steinlechner), "the fat and fanatic South-Tyrolese with the penetrating brown eyes and the large nose" and little Eric suffered already under the nervous

---

[60] During this time Karl von Terzaghi tried to receive a professorship at a British university (see Glossop, 1973).

tensions between both women. Karl dismissed Mitzi immediately with a generous severance pay.

Furthermore, owing to wild rumors about the Germans marching into Czechoslovakia, he decided therefore to bring Eric immediately to Paris. During these exciting days his staff mechanic, Sentall, was a great help for Karl, e.g. he took care of bringing documents in order and of getting visas. Moreover, he promised to arrange for toddler Eric to ride with the nursemaid Erna to Mannheim. There, Mary, wife of O.K. Fröhlich, should take charge of him and bring him to Paris.

On May 23, Karl rode to Paris without trouble at the border. In a letter dated the same day, he informed the Rector that he would go to Paris and London for some consulting work, and that his co-worker Kienzle would substitute for him in his lectures. Ruth and Karl von Terzaghi got very nervous concerning Eric's trip to Paris, above all due to O.K. Fröhlich's telegram which was unclearly formulated. However, on May 28, Mary Fröhlich arrived safely with Eric in Paris and the von Terzaghi family was reunited. The von Terzaghis took up quarters in a boarding house in St. Germain, and one week later they moved to "a cosy old house, high and narrow, with mansard and an old garden completely overgrown with ivy." This Louis XIV house became their temporary home.

Karl returned with the Orient Express to Vienna on June 9/10. Several days later he had a talk with counsellor Etscheit who told him that "he had smoked a cigarette with an important man. My case is not hopeless. Payment, 1,000 Reichsmark immediately, 2,000 after settlement."

In the middle of June, von Terzaghi finished his work on the opinion paper and, on June 27, he gave his last lecture to his Viennese students, ending his academic career at the Technische Hochschule.

At that time the discontent with the Nazis was growing. In particular, the treatment of Jewish fellow-citizens bothered Karl von Terzaghi: "Jews are released at 3 h, at 5 h again arrested. New law in preparation that Jews must deliver their complete means abroad. In Eisenstadt wholesaler Wolff, family since 15th century in Eisenstadt. Funded museum, custodian by himself, known philanthropist. Completely expropriated, must be outdoors as beggar. Safety of the Jews not guaranteed in streets of Eisenstadt."

In front of the Technische Hochschule Karl von Terzaghi saw a group of people and in the middle Jewish students squatting on the ground and cleaning the sidewalk.

At the end of June he visited Ella in Graz and complained about the new political situation in Austria. "The new system is unbelievable proletarian and banal, the treatement of the Jews brutal."

In the next days, he sorted out his files in Vienna. In the evening of July 8, he visited Mitzi (Maria Obermayer). They were sitting in the old fashioned room which he had occupied when he was on vacation in Vienna before 1925. A melancholy mood lay over the evening; it was his last evening in Vienna.

On July 9, 1938, he was, via Mannheim and Heidelberg, on his way to Paris. In his diary, he drew a line. Obviously, it was clear to him that he would never again return to Vienna. His co-workers, his colleagues, and even some relatives in Graz had not been informed; nobody knew exactly where von Terzaghi had gone.

In July and August, 1938, von Terzaghi became seriously ill, but he was able to recover by the sea. After that, he travelled to London, returning to Paris at the end of August. In letters to the ministry for education and the Rector of the Viennese Technische Hochschule dated August 31 and November 24, 1938, von Terzaghi urgently requested that he be dismissed at the end of the winter semester 1938/1939, and that he be suspended from office immediately, without salary, due to his bad state of health after twenty years of continuous teaching and research work. He pointed out that, in 1929, he had accepted the professorship at the Viennese Technische Hochschule despite a loss in personal financial earnings, although he could have accepted American citizenship after having worked such a long time in American duty. Moreover, in this connection, he also spoke of the fact that certain promises in the negotiations with the administration had not been fulfilled so that he had been obliged to cover the expenses of his chair partially with his own funds. Furthermore, he stated: "In 1936, my colleague Professor Fillunger started the attempt to destroy my good reputation as a human being and a researcher. And the prevention of this attack hindered me for half a year in continuing my scientific work." He concluded that the work conditions at the Technische Hochschule had been for him by no means ideal. Von Terzaghi remarked: "I had, however, the satisfaction of holding off the non-Aryan elements from my chair and my conscientious training of a larger number of talented students to provide my nation with remarkable services. Due to my open preference for the nationalist-minded part of the academic youth, I had to renounce support for my chair from the old government for years. The employees who have come from my school have found in the German Reich a highly-deserved appreciation." These naive remarks show that he was basically politically uninvolved, although he had spent influential years of his life in the heavy political surroundings of Europe. Among his circle of friends in Vienna were some *Nazis.* It seems, however, that Karl von Terzaghhi considered these friends only as acquaintances. His departure from Vienna, giving up his position and rights at the Technische Hochschule, was also not politically motivated, as some rumors had implied. It was rather his restless nature that led him to seek a new assignment. His flattering response – among other things, he signed now his letters with *Heil Hitler* – to the new leadership of the Technische Hochschule after the annexation in 1938, is obviously founded in his great vanity, which he also showed in situations other than in his political view.

Moreover, von Terzaghi pointed out that, as a result of the annexation in March, Mrs. von Terzaghi was forced either to renounce her American citizenship and her freedom of movement, or leave the country. In view of the premises under which she had decided at that time to move to Europe, it was understandable that she refused to stay in Germany. She was forced to take the necessary steps that the new political situation demanded. Von Terzaghi closed his letter with a request: "I request in view of the described circumstances to offer such a resignation from the Technische Hochschule of Vienna that my personal and scientific relations to my old co-workers suffer no damages."

The temporary Rector Professor Saliger answered von Terzaghi in a letter dated September 6, in which he expressed his disappointment over von Terzaghi's request; he reported to the Minister on September 7. In this letter, the Rector asked the Minister to refuse von Terzaghi's request concerning his suspension, and to dismiss him immediately. Saliger was, in particular, so disappointed because von Terzaghi had given to him his word of honor to come back from London to Vienna and he broke it.

In the middle of September, von Terzaghi returned to London and, on the evening of September 16, 1938, he left Europe again, bound for the United States.

In order to clean up the affair after von Terzaghi's departure, the Rector of the Technische Hochschule established a committee which consisted, among others, of the professors Saliger, Hartmann, Schaffernak, Kozeny, and *Dozentenführer* Schober (the Nazis had installed someone in the position of *Dozentenführer* at every major university). In a letter, the committee summarized the results of a meeting held on September 28, stating that the chair (Water Construction II) had not been handed over in an orderly fashion by von Terzaghi. Furthermore, it was stated that von Terzaghi had moved to Paris in July, 1938.

Moreover, the idea of suspending him from his duties during the winter semester 1938/39 and replacing him in his seminars with his co-worker Dr. Kienzle was refused. The committee proposed that Professor Schoklitsch, a former professor at Brno, should replace von Terzaghi. Finally, the committee painfully accepted von Terzaghi's resignation and asked him to reconsider his decision because, at that time, huge tasks in the building field which necessitated his cooperation were forthcoming. In a letter dated October 27, 1938, the Minister of Education refused von Terzaghi's request concerning suspension. This decision was conferred to von Terzaghi by the Rector of the Technische Hochschule. In a telegram from Cambridge, USA, of November 25, von Terzaghi requested immediate dismissal from his duties.

In the time to follow, some strange things happened in the wake of von Terzaghi's dismissal. Obviously, von Terzaghi was very much interested in weakening the negative image surrounding his obscure disappearance from Vienna, for he was also very much interested in receiving the thanks of the Führer in his letter of dismissal. In two nearly identically formulated letters written on April 3, 1939, one to von Terzaghi's daughter, Vera, and the other to his wife, the Rector stated that the thanks of the Führer and Reichskanzler for von Terzaghi's achievements could only be included in the letter of dismissal if the *Arier-Nachweis* (evidence of pure Aryan heritage) for von Terzaghi and his wife was successful. Mrs. von Terzaghi replied in a letter of April 12, 1939, pointing out that it would be extremely difficult to obtain the required documents as a result of the fact that in the United States it was not usual to keep all the documents. Von Terzaghi himself wrote a letter to the Rector on April 24, 1939, in which he pointed out that he had received a copy of the letter to his wife. He stated that it would be easy for his brother-in-law, Professor Dr. Byloff (Graz, Austria) to completely document his lineage. Before his marriage, he had naturally been interested in his wife's family tree. However, from memory he was only able to give a little information regarding the history of his wife's family. Finally, he made some general remarks referring to his wife's family history: "In the United States the social separation between Aryan and non-Aryan elements is essentially sharper than in Germany and Austria. The circle of friends of my wife's family consists solely of Aryans, and the whole personality of my wife is to such an extent Aryan, that the desired proof concerning the descent of my wife's origins seems to be a mere formality. I do not have any doubt concerning the result."

On April 26, 1939, the Rector of the Technische Hochschule requested, in a letter to Professor Dr. Byloff, that he send him the descent documents. At the end of the letter he stated: "The proof of Aryan descent is necessary for the dismissal-degree of Professor Terzaghi." In letters dated April 27 and August 8, 1939, the Rector reported to the minister on his efforts concerning the proof of Aryan descent for von Terzaghi and his wife, saying that his efforts had not been successful. Finally, he applied for von Terzaghi's dismissal without the thanks of the Führer. In a letter of November 18, 1939, the Rector declared to von Terzaghi that a dismissal could only be effected if he would renounce his salary and his pensions. Von Terzaghi agreed to this proposal on December 8, 1939. Moreover, he requested that the letter of dismissal be sent to his address in the United States.

The *Ministerium für innere und kulturelle Angelegenheiten Abtl. IV, Erziehung, Kultur und Volksbildung* in Berlin wrote to the Rector on November 4, 1939, to dismiss von Terzaghi. The dismissal was valid from the end of March 1939. Moreover, in the same letter, it was left to the Rector as to whether to express thanks.

The letter of dismissal (dated September 22, 1939), as well as the corresponding document, did not contain any words of thanks, which Karl von Terzaghhi desired so much.

After this, several rumors circulated about von Terzaghi's departure from Vienna. In a letter of August 31, 1943, the Rector reported to the *Reichsminister für Wissenschaft, Erziehung und Volksbildung* in Berlin, saying that von Terzaghi had moved to the United States without an understandable rea-

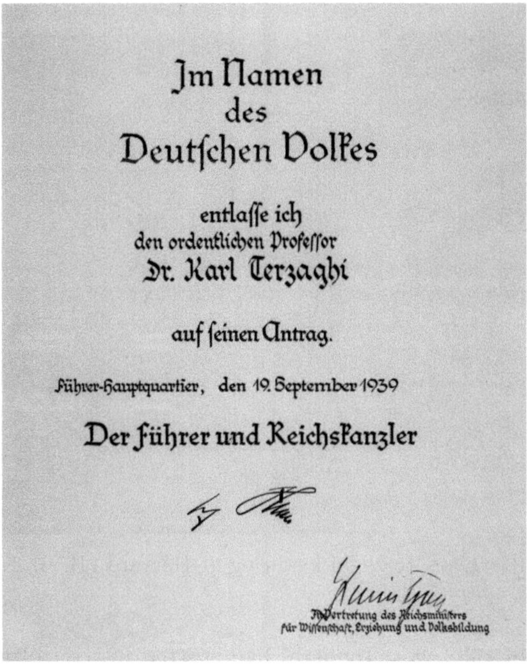

Im Namen
des
Deutſchen Volkes

entlaſſe ich
den ordentlichen Profeſſor
Dr. Karl Terzaghi

auf ſeinen Antrag.

Führer-Hauptquartier, den 19. September 1939

Der Führer und Reichskanzler

In Vertretung des Reichsminiſters
für Wiſſenſchaft, Erziehung und Volksbildung

Dismissal document.

son. Therefore, he continued, the name von Terzaghi should not be mentioned in the correspondence of the Hochschule. Furthermore, the Rector declined a personal ceremony on the occasion of von Terzaghi's sixtieth birthday. Some of von Terzaghi's colleagues were also deeply disappointed by his sudden and somewhat strange disappearance. In his memoir, Professor Salinger wrote: "He (von Terzaghi) has stealthily left Vienna and Austria soon after their incorporation into the Reich and our trust has been coarsely deceived."

Von Terzaghi arrived in America on a clear day on September 25, 1938, at 7.30 a.m. He was welcomed by A. Casagrande. His wife and son had remained in Paris. The next day, von Terzaghi had a meeting with Dean Westergaard of Harvard University, obviously in order to discuss his future position at Harvard. A. Casagrande had already made great efforts to find a position for von Terzaghi at a university in the United States in the beginning of 1938. He had written, for example, to Dean S.B. Moris at the School of Engineering, Stanford University, California, asking for a half-time professorship for von Terzaghi. However, Dean Morris replied by letter: "Our budgetary situation does not permit me to make any encouraging statement at this time ..." Finally, von Terzaghi acquired a position as a visiting lecturer at Harvard University, though with a smaller salary and, in the semesters to follow, he lectured on Engineering Geology and Applied Soil Mechanics.

Karl von Terzaghi lecturing at Harvard (1950).

By the end of 1938, von Terzaghi had started his consulting work in the United States and this was to intensify in the following years. He began as a consultant for the Department of Subways and Traction of the city of Chicago, in connection with the construction of a subway system and, in the next two years, his activities were focused mainly on this project. In April, 1939, he returned to London to lecture at Imperial College. Before leaving for London, he was asked to investigate the considerably large settlement of the Charity Hospital in New Orleans.

In the spring of 1939, von Terzaghi's wife and son moved from Paris to Cambridge and, in June, 1939, von Terzaghi and his family settled in Winchester, Massachusetts, a small town to the north of Cambridge, first in a rented house; later the von Terzaghis bought a piece of land in Winchester and built a new one. The von Terzaghis enjoyed their home and the beautiful surroundings on the edge of Mystic Lake. In addition, they were a very hospitable family. They frequently lodged colleagues and scientists from the US and abroad.

In May, 1941, little Margaret (Peggy) was born and in March, 1943, Karl von Terzaghi became an American citizen.

As has already been mentioned, from December 1938 on von Terzaghi was involved primarily in the Chicago subway project. Although most of his previous practical application of soil mechanics had been in connection with foundations and dams, he did not hesitate to take on the new task. In January, 1939, von Terzaghi arrived for the first time in Chicago in connection with the subway project, and then followed this during the next three years

Karl von Terzaghi on a building site.

Karl von Terzaghi with Ralph Peck.

with many extensive advisory visits, which became less frequent during the last year. In 1942, he was asked to investigate the stability of the Necaxa Dam in eastern Mexico. During the following years, von Terzaghi acted as a consultant to the Mexican Government, giving advice concerning the further survival of the City of Mexico due to the lowering of the water table. Apart from these projects, he also found time to write two remarkable books on *Theoretical Soil Mechanics* (von Terzaghi,1943) and Soil Mechanics in Engineering Practice (Terzaghi and Peck, 1948), which attracted much interest throughout the world and were translated into many different languages. The most interesting book – from the scientific point of view – was the first one. In the Preface of this book von Terzaghi pointed out: "In the fifteen years since the author published his first book on soil mechanics interest in this subject has spread over the whole globe and both our theoretical and our practical knowledge of the subject have expanded rapidly. The *Proceedings of the First International Conference on Soil Mechanics* (Cambridge 1936) alone contains a greater amount of quantitative information regarding soils and foundations than the entire engineering literature prior to 1910. Yet, as in every other field of engineering, the first presentation of the theoretical principles has been followed by a period of transition characterized by a tendency toward indiscriminate application of theory and by unwarranted generalizations. Hence, when the author began work on a new textbook on soil mechanics he considered it advisable to separate theory completely from practical application. This volume deals exclusively with the theoretical principles.

Karl von Terzaghi in action, Rotterdam (1948)

Karl von Terzaghi and Professor Veder at a building site in Switzerland in 1953

Theoretical soil mechanics is one of the many divisions of applied mechanics. In every field of applied mechanics the investigator operates with ideal materials only. The theories of reinforced concrete, for instance, do not deal with real reinforced concrete. They operate with an ideal material, whose assumed properties have been derived from those of the real reinforced concrete by a process of radical simplification. This statement also applies to every theory of soil behavior. The magnitude of the difference between the performance of real soils under field conditions and the performance predicted on the basis of theory can only be ascertained by field experience. The contents of this volume have been limited to theories which have stood the test of experience and which are applicable, under certain conditions and restrictions, to the approximate solutions of practical problems ...

For the author, theoretical soil mechanics never was an end in itself. Most of these efforts have been devoted to the digest of field experiences and to the development of the technique of the application of our knowledge of the physical properties of soils to practical problems. Even his theoretical investigations have been made exclusively for the purpose of clarifying some practical issues. Therefore, this book conspicuously lacks the qualities which the author admires in the works of competent specialists in the general field of applied mechanics. Nevertheless, he could not evade the task of writing

Karl von Terzaghi with the
brilliant young Lauritz Bjerrum, 1958.

the book himself, because it required his own practical background to assign to each theory its proper place in the entire system ..."

Although von Terzaghi's book carried the title *Theoretical Soil Mechanics* the subject was far removed from a closed theory. Soils, like sand and clay, were treated as one-component continua whereby the treatment of their mechanics revealed that he was not familiar with continuum mechanics. Moreover, nearly half of the book is devoted to the failure of soils and surprisingly to a great extent to the old earth pressure theories of Coulomb and Rankine, which he considered as antiquated in the beginning of the 1920s. The chapter on the mechanical effect of water in soils is meager. Although von Terzaghi (1925b) had recognized long before that a water-saturated soil body was a mixture, he did not use the mixture theory or elements of it; in no place in his work did he separate the constituents porous soil (skeleton) and water (liquid) considering all the interaction effects between the constituents, as Fillunger (1936) had done. Thus, he repeated his treatises of the 1920s and 1930s on the permeability and the consolidation of soils. The chapter on elasticity in soil mechanics was based on the linear elasticity and can be passed over in silence.

The basic shortcoming of the book is that it lacked a clear structure as is usual in continuum mechanics and contained too many examples whose

The President at the opening session (Fourth International Conference, London, 1957). Karl von Terzaghi first and A. W. Skempton second from the right

theoretical treatment could not be duplicated in many cases. Moreover, there were many misprints and the section on bearing capacity was not correct as Karl von Terzaghi conceded in a letter of November 17, 1952, to Dr. G. G. Meyerhof, of England, who had already sent him a long list concerning misprints in *Theoretical Soil Mechanics* in 1944. Unfortunately this book has influenced the scientific way of working in soil mechanics to a great extent.

In the years to follow, von Terzaghi extended his consulting activities. In particular, he gave advice on foundation problems as well as on water and power projects in British Columbia, Canada. However, he was also a welcomed consultant in Sweden, Brazil, Turkey, and France. (Karl von Terzaghi's consulting work is extensively described by Goodman, 1999). Sometimes he connected his long trips with visits to relatives or friends, e.g. in October, 1948, he spent a week with his lifetime friend Hans Kalbacher in Montevideo, Uruguay.

How extensive his consulting and lecturing work was can be read from a letter to his daughter Vera in Austria, which he had begun in the Westward Hotel in Anchorage, Alaska. In this letter, he described his activities between July and November, 1952, at which time he was nearly 69 years old. At the beginning of July, 1952, he started his third trip to the northwest of the

Karl von Terzaghi at home

Karl von Terzaghi travelling in Cuba

United States. First, he went to Urbana, south of Chicago, to lecture at the university there. From Urbana, he proceeded to Spokane, Washington, from which he visited, by ship and plane, the tremendous landslides which had occurred on the northern part of the Roosevelt Sea, destroying the roads on the plateau. In Trail, on the Columbia River, north of the Canadian border, he visited the drainage tunnel which he had designed. In Vancouver, British Columbia, he conferred with the directors of the electric plants for three days and gave a talk to the Canadian Engineering Society. From Vancouver, von Terzaghi flew with a small plane to the Nechako River, west of Prince George, in order to see the building site of a rock-filled dam, measuring one hundred meters high designed by himself. One week later, he visited the building sites of other small dams by plane. Again by plane, he flew to the west, across the coastal mountain range, to the building site of a power station, and visited the line of construction for a long-distance line in the glacial coastal mountain range. The terrain was so inaccessible that the various points could only be reached by helicopter. He returned to Vancouver and admired the grandious glacial fjord landscape. In Vancouver, two telegrams awaited him, the first one calling him back to Trail, the second one requesting him to fly to Southern California. He travelled back to Trail, on the Columbia River, and then went by plane to Los Angeles, then to friends in picturesque La Jolla, where he spent a nice weekend. From California, he returned to the east, to Hartford, Connecticut, where he visited a dam. After spending one week on vacation in Maine with his family, von Terzaghi was again on his way to Chicago, where he had to give the opening speech on the occasion of the one-hundredth anniversary of the American Society of Engineers. In the south of Chicago he also had to review an underground storeroom. He returned from Michigan to Winchester, hoping to be able to relax at his home. However, three days later, a telegram was calling him to Alaska. On his birthday, he flew again to Vancouver and Trail, and then back to Winchester. At the end of October, he lectured again in Urbana and, at the end of November, he was on his way to Mexico, where he received an Honoray Doctoral Degree at the University of Mexico.

It is amazing that von Terzaghi could still physically and mentally stand such exhausting trips at the age of 69.

In 1953, von Terzaghi suffered a heart attack. After a stay in the hospital, he recovered fully. A few months later, he studied the Sasumusa dam site in the Kenya Colony in Central Africa. In 1954, von Terzaghi was asked to take over the position of the Chairman of the Board of Consultants for the planned High Aswan Dam in Aswan, Egypt. From then on, he took the chair in numerous meetings of this board. Gradually, his eyesight became weaker and in 1958 he had "to go to the hospital for a cataract operation." In 1959,

Karl von Terzaghi, 1958.

Karl von Terzaghi on his 75th birthday

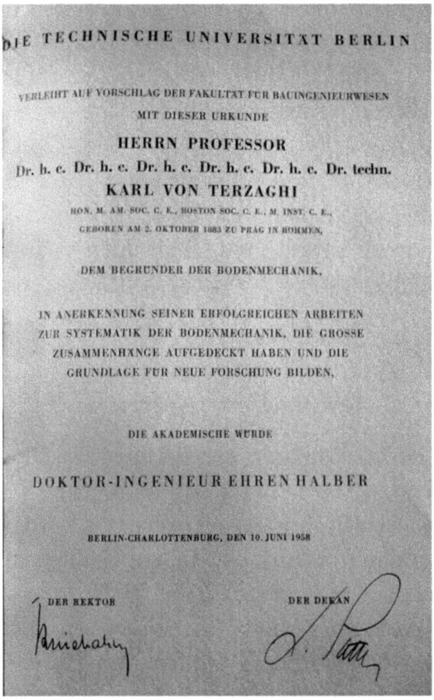

DIE TECHNISCHE UNIVERSITÄT BERLIN

VERLEIHT AUF VORSCHLAG DER FAKULTÄT FÜR BAUINGENIEURWESEN
MIT DIESER URKUNDE

HERRN PROFESSOR
Dr. h. c. Dr. h. c. Dr. h. c. Dr. h. c. Dr. h. c. Dr. techn.
KARL VON TERZAGHI
HON. M. AM. SOC. C. E., BOSTON SOC. C. E., M. INST. C. E.,
GEBOREN AM 2. OKTOBER 1883 ZU PRAG IN BÖHMEN,

DEM BEGRÜNDER DER BODENMECHANIK,

IN ANERKENNUNG SEINER ERFOLGREICHEN ARBEITEN
ZUR SYSTEMATIK DER BODENMECHANIK, DIE GROSSE
ZUSAMMENHÄNGE AUFGEDECKT HABEN UND DIE
GRUNDLAGE FÜR NEUE FORSCHUNG BILDEN,

DIE AKADEMISCHE WÜRDE

DOKTOR-INGENIEUR EHREN HALBER

BERLIN-CHARLOTTENBURG, DEN 10. JUNI 1958

DER REKTOR

DER DEKAN

Document of the TH Berlin
on the honorary doctor

the Egyptian government decided, due to disputes with the western countries, to carry out the project with the financial and technical assistance of the Soviet Union. Von Terzaghi resigned immediately because there was now no possibility to make any influence on the design and the realization of this enormous engineering project.

In his scientific and engineering career, he received numerous honors. Among others he was a member of the Österreichische Betonverein, the Academy of Science in Vienna, the American Society of Civil Engineers, and the Boston Society of Civil Engineers. He received the Norman Medal of the American Society of Civil Engineers four times. As well as other awards, he was also honored by the "Goldene Ehrenmünze" of *The Österreichische Ingenieur- und Architekten- Verein.* His outstanding achievements were recognized worldwide as he was bestowed with several honorary doctor degrees. At the age of 65 years, he received his first "Dr. sc.h.c." by Trinity College of Dublin. In the years to follow, he was awarded with additional honorary doctor degrees from universities in Istanbul, Mexico D.F., Zürich, Bethlehem (USA), West Berlin, Trondheim, Graz, and Columbus, Ohio (USA).

His last years were overshadowed by a serious operation in 1960 and the subsequent loss of one leg. However, with his iron willpower, he was able to continue his work.

But gradually he recognized that his time was running out. He formulated his last will in the form of a letter to his wife:

"October 2, 1961(?)

Wwx (Ruth, the author)

In the event of my death:

My body should be cremated, and no services or rites should be held on account of my profound dislike for ceremonies of this kind. My ashes should be buried in South Waterford where I have spent with you many happy days, and a simple tablet should be erected with the following inscription:

Karl Terzaghi,
a civil engineer
born October 2, 1883, in *Prag*, Austria and died ...

He has lived without compromising, served his chosen profession to the best of his abilities and died without anything to regret.

After my departure Bjerrum should be invited to come to Cambridge, with all expenses to be paid out of my estate and collect the items which I have selected for the T. library in Oslo. During his sojurn you should invite him and our intimate friends to a party to rejoice on the fact that I was granted and you have shared, a long and wholesome life which I was able to live without humiliating compromises, in accordance with my innate pattern. There is nothing to feel sad about it, because the end is an essential and inevitable part of the existence of our species.

Lovingly yours
Bear[61]."

In the last weeks before his death on October 25, 1963, this excellent and experienced civil engineer and great consultant wrote several farewell letters to old friends, e.g. his staff mechanician Sentall and the artist Hilde Uray, though none to his close disciples, and he read in his diaries, calling to mind his successful and bizarre life, which was likely formed by his Italian and Bohemian heritage. And he recognized that the end was near, an end with no future, no hope, no return.

[61] Karl von Terzaghi's nickname.

5.2 Some Developments in Soil Mechanics

Karl von Terzaghi's negative attitude towards the scientific penetration of soil mechanics and his turn to practice (consulting) and the observational method has strongly influenced many engineers. This method and experience stand, of course, at the beginning of every field in natural and engineering science. Karl von Terzaghi and his followers, in particular, Ortenblad (1930), Biot (1935) and Carrillo (1942) failed to put soil mechanics in a scientifically founded theory. The a.m. authors tried to transfer von Terzaghi's partial differential equation for the one-dimensional motion to the three-dimensional case, without a profound mechanical foundation and thus constructed theories which neglected important physical phenomena. Moreover, their attempts have hindered the theoretical development of soil mechanics for a considerable time. Furthermore, many engineers, although occupied at technical universities and other scientific institutions, eagerly followed von Terzaghi's view of the observational method, avoiding every scientific investigation while working on the side as consultants.

The main efforts were on the contrary directed towards carrying out the ideas on soil mechanics which had been developed by von Terzaghi and his followers. At the beginning of the 1930s, the idea was born to establish National Committees in soil mechanics and foundations. In 1935, von Terzaghi

Arthur Casagrande

stated in a letter to A. Casagrande: "I immediately took a look at your draft concerning the soil conference in Cambridge ... At first glance I only miss the step from the individual to the organization. The current draft contains only an 'Application Form' for the individual participants. If I understand it in the right way, the procedure is as follows: at first the national-committees are nominated and only then will the pamphlet be mailed so that the names of the members of the national-committees are already contained in the pamphlet. It would be doubtful as to which way the national-committees should be called to commence. I believe in different countries there should be different procedures."

On October 9, 1935, A. Casagrande answered. He proposed that von Terzaghi should decide the way in which to form the national-committees in Austria, Sweden, France, and Russia. For the remaining countries – in particular, Germany, Italy, Egypt, etc. – he himself would arrange the formations. The efforts to establish national-committees continued with much success in the years following. Finally, A. Casagrande stated in a letter to von Terzaghi of February 1, 1938: "I would like to congratulate you on the outstanding success of your efforts for the organization of national-committees. – Thus we can consider the following countries as settled: England, France, Italy, Switzerland, Austria, USA, Mexico, China, Egypt, Russia. In the following countries we can be assured that, if we make an effort soon, an organization could be set up without difficulties because there are qualified men: The Netherlands, Denmark, Sweden, Finland, Estonia (perhaps one could found a Nordic committee in which Denmark, Sweden, Norway, Estonia and Finland act?), Hungary, The Dutch Indies. In addition, probably Canada, Brazil, Australia, Poland. Thus only the complicated cases remain, Germany and Czechoslovakia, and some countries like Turkey, South Africa, New Zealand, Palastine etc.

The next necessary step will be to nominate the Executive Committee of the Permanent International Organization. According to the Resolution No.3, this Committee should consist of three members, besides you and me. For these three members I suggest: Proctor, Bretting, Meyer-Peter or Buisman. Germany is represented indirectly by you and I, however, not officially so that France and England cannot be upset about this.

The international committee (point 'd', Resolution No.3) should be elected by the individual national-committees, and the question arises as to how many members each country is allowed to elect or should elect."

The above-mentioned resolution, No. 3, was adopted by the participants of the Harvard conference after some discussions at the end of the last meeting on June 25, 1936.

In the letter to von Terzaghi from February 1, 1938, A. Casagrande announced further steps: "The next step will be the formation of an International Society for Soil Mechanics. The membership fee should be kept low in order to allow all seriously interested people the membership. The most suitable use of the fees is in my opinion for a quarterly 'Soil Mechanics Review' in several languages. I suggest that either you in Vienna, or I here in Cambridge should do the editorial work and that one should gain, for each of the most important languages, one co-worker who has a good knowledge of soil mechanics and to whom the corresponding literature is easily accessible. He would briefly summarize the most important papers in this language or oblige the corresponding author to write the summary by himself. These contributions would be sent to the editorial board, further shortened if necessary, sorted and put together for one issue and then sent to the individual editorial boards, which would take care of the translation and the print. For printing, one should choose a cheap and quick process (like our Proceedings, for small numbers of copies *mimeographed* [62]). At the beginning, two or three languages will be sufficient. Over the course of time I believe that English, French, German, Spanish and Swedish (or Danish) editions will appear. The publication of the English edition could be taken over by the local center of the International Society, and the remaining editions will have to be managed by the national-committees. I suggest financing this in such a way that each member of the International Society would have the right to receive the Review in one of the languages for a price of, e. g., $ 3.00 per year, whereas for non-members the price, would for example be the double. The edition in a new language would only then be undertaken when a sufficient number of members had declared their intention to subscribe to it. If necessary, the price could be at different levels for different languages. One could fix the membership fee at $ 1.00. The amount which can be saved can be used as a contribution to the publication of the proceedings of the next conference.

I would be grateful to you if you would think over all of these problems and would report to me proposals for changes and supplements. As we have agreed I will send an extended printed circular letter to all members of the first conference which will contain all this, as well as the regulations of already existing national-committees and membership (as a model and stimulation for other countries). In this treatise, I will ask the members to comment on the proposals in order to give the Executive Committee the opportunity to consider the opinions of the members before the final organization of the International Society is carried out."

In order to publicize and make known their standpoint on soil mechanics, the von Terzaghi-school came up with another idea: to write books on the subject. In a letter of November 3, 1936, written to the von Terzaghi family, A. Casagrande stated: "Some days ago, Mr. Hamilton[63] was here and told me of the 'goodbyes' in New York. Among other things, I have complained to him about the new parasitism in soil mechanics, whereupon he correctly remarked: 'It is important that we edit our books.' This is probably the only solution." Indeed, as is well-known in soil mechanics, as aforementioned von Terzaghi

[62] A special kind of typography.
[63] From the publishing house John Wiley & Sons.

wrote two remarkable books on theoretical and practical soil mechanics in 1943 and 1948, the last one with Professor Peck, a former co-worker of his.

Moreover, the von Terzaghi-school did not hesitate to revile other scientists who did not share their views on soil mechanics. For example, in 1936, Professor Housel from the University of Michigan, an already-recognized scientist in the field of soil mechanics, had his own ideas which did not coincide with those of the von Terzaghi-school. At the 1936 conference, the reports over his extensive work were restricted by the organizing committee to a summary of less than five pages, and his talk to ten minutes. Housel complained bitterly about this situation:

"Ten minutes is far too short a time to present completely one's position on one or several controversial points which have been the subject of discussion during the meetings of this Conference. Consequently I have attempted to formulate as briefly as possible in written form some of my own reactions which have been accumulating during the discussion of the past several days and which are clamoring for expression too insistently to be ignored for my own peace of mind." Housel spent much of his time after the 1936 conference attempting to discredit various of von Terzaghi's concepts.

Von Terzaghi did not react very much to these attacks. It was A. Casagrande who carried the banner higher than the leader. With ironic and sometimes rough words, he attacked several researchers: In 1936, for example he wrote: "A propos, Mr. Knappen has also visited me and has told me with pride that he is the Secretary of the Executive Committee of the new division of the Am. Soc. Civ. Eng. If one hears him talk one would assume that he has eaten soil mechanics with a ladle. He and Philippe are writing a book together! The new division works as I have expected. Now they fortunately also have a committee on the 'Classification of Soils, Terms and Definitions in Soil Mechanics' with W.P. Kimball, our common good friend, as the only chairman. He is of course very much qualified for that because he is one of those who will standardize everything in his stupidity. I could kick myself that I allowed myself to be persuaded to be a member of this Society."

Professor Burmister, a famous professor at Columbia, was another person who was adversely affected. In a letter to von Terzaghi from May 31, 1937, A. Casagrande pointed out: "I believe that neither Hogentogler's nor Plummer's book will cause much damage because they are written in such a manner that anyone who has some intellect must see how useless these books are. The more dangerous is in my opinion the book which Burmister will publish with Wiley (!). I have expressed without reserve my opinion of Burmister to Hamilton. However, Burmister has a lot of 'pull' in New York through his colleagues at Columbia, who support him greatly and who consider him as a man of genius, as well as from the New York Consulting Engineers for whom he works so much. It is really a pity that such a big-mouthed braggart has spread himself so far in New York. Such valueless works as Burmister and Krynine have published in the Proceedings A.S.C.E. in the last months have been accepted without further ceremony by the publication committee, whereas an outstanding work which Taylor of M.I.T. submitted, was refused. It would be really interesting to know who the specialists are that the publication committee seeks for advice. Under such circumstances it will, of course, never occur to me to publish something in the Proceedings. One cannot dare by any means to write an honest criticism of Krynine's work, because they would never publish it, and thus confess what an idiocy they have taken on."

A. Casagrande was not the only follower who tried to carry the Harvard point, although he was the leading person among the other three important adherents, namely, the professors R. Peck from the University of Illinois, Urbane, Illinois, A.W. Skempton from the Imperial College of Science and Technology, London, and L. Bjerrum, Director of the Norwegian Geotechnical Institute, Oslo. Their efforts to reveal von Terzaghi's dominance in the field of soil mechanics culminated in the publication of the book *From Theory to Practice in Soil Mechanics* (Bjerrum, Casagrande, Peck, Skempton, 1960), intended to serve as an anniversary volume for Karl von Terzaghi's 75th birthday. However, as is usual for such commemorative publications with a fixed date, the publishing did not occur in 1958 but in 1960. In the preface, the four disciples described the purpose of the book: "Its purpose is to present a detailed account of Terzaghi's personality, his professional achievements, and his method of working. To the greatest possible extent the essence of his contributions is presented in his own words. When set forth in orderly sequence and supplemented by explanatory comments, Terzaghi's writings and reports show clearly how he proceeded, step by step, first to establish the fundamental principles of soil mechanics and then to use them as powerful tools in his engineering practice. Hence the contents of this book provide an ideal means to demonstrate, especially to young engineers, the prerequisites and techniques for successfully practicing soil mechanics. This belief was not only one of the incentives for the preparation of the book but it was also responsible for the selection of the title."

The book consists of five major chapters: Life and Achievements; Aim, Scope, and Methods in Soil Mechanics; Selected Professional Papers; Selected Professional Reports; Bibliography. The preparation of the manuscript did not run without difficulties and some strange occurrences. It had been intended by the editors to publish the anniversary volume with the publishing house John Wiley & Sons, Inc. However, it was not sure for a considerable time whether Wiley would agree. In March, 1959, A. Casagrande wrote to Skempton: "..., Peck is confident that Wiley will accept publication just as soon as they see the assembled manuscript. But I am certainly glad to know that the prospects would also be good in England, in case we get bogged down here." Finally, however, Wiley published the anniversary volume.

There were also other reasons holding up the publishing at the appointed date. We have already mentioned the struggle surrounding Fillunger's contribution to the concept of effective stresses and Karl von Terzaghi's correct understanding of this principle. A. Casagrande remarked bitterly to Skempton that he had failed in his article on the effective stress principle to give von Terzaghi the sole credit for the formulation of this principle: "I am not happy about the last two paragraphs of your article, for two reasons. First, because I cannot help but feel that many readers will receive the impression that Terzaghi has done the spade work, and has done it well, but then he missed the boat and let somebody else discover the crowning achievement; and that, therefore, the crown belongs to Rendulic." A. Casagrande's second objection concerns some details of the constitutive behavior of clay. Also the inclusion of parts of von Terzaghi's first book *Erdbaumechanik* was not undisputed. Bjerrum stated in a letter to Peck of June 3, 1958: "As mentioned in previous letters Terzaghi's way of arranging the material is very complicated and most of the book is difficult to understand. Indeed, you need to be extremely familiar with the material in order to appreciate his ingenuity."

Also the interference of Karl von Terzaghi himself had led to a delay. This interference ranks as the strangest occurrence during the preparation of the book. A. Casagrande had undertaken the task of writing Karl von Terzaghi's biography which appeared in the volume under the title *Karl Terzaghi – his life and achievements by Arthur Casagrande*. However, in a letter to Bjerrum of September 9, 1958, and only to Bjerrum, von Terzaghi maintained that he had written the biography by himself: "(The biography) was written by A.C. It read like an enclosure to an application for a job. I scrapped it and wrote it myself (40 handwritten pages). Now it seems to be juicy and lively."

A. Casagrande did not say any word about this embarrassment. On March 7, 1959, he outlined in a letter to Skempton: "I hope to be able to reciprocate soon with the biography. It is now in the hands of Terzaghi for a final review."

In the late 1940s and early 1950s, it was recognized by several scientists that the Harvard-school was dominating soil mechanics and that the views of other researchers were put down by von Terzaghi and his disciples, above all by A. Casagrande. These scientists were concerned that the Harvard-school was dominating the soil mechanics profession, especially A.S.C.E. (American Society of Civil Engineers), and controlling the publication of papers. Moreover, it was rumored that Professor Taylor of M.I.T. was never promoted to full professor allegedly because of von Terzaghi's intervention. It was also rumored that von Terzaghi tried to have Professor Tschebotarioff fired from Princeton. The story was always the same. In order to be successful, one had to have attended Harvard and had to be obedient to von Terzaghi's views. It was then believed that every paper in soil mechanics submitted to the A.S.C.E. had to have von Terzaghi's approval to be published. However, one must state that there is no evidence that von Terzaghi personally prevented publication. On the other hand, the late Professor A. Casagrande was instigating the censorship in von Terzaghi's name. He suppressed several outstanding pieces of work.

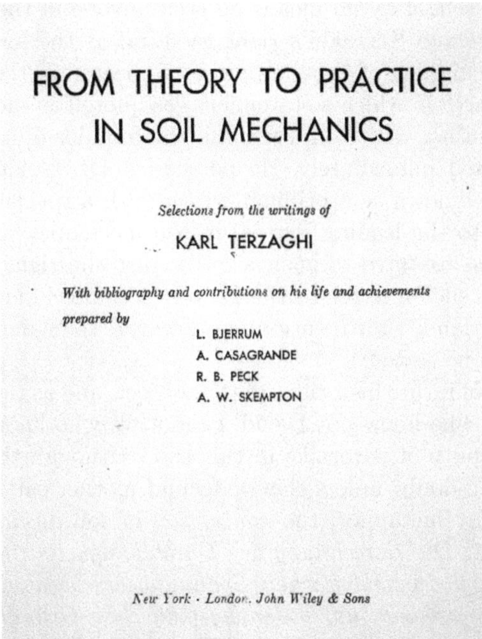

FROM THEORY TO PRACTICE
IN SOIL MECHANICS

Selections from the writings of

KARL TERZAGHI

With bibliography and contributions on his life and achievements prepared by

L. BJERRUM
A. CASAGRANDE
R. B. PECK
A. W. SKEMPTON

New York · London. John Wiley & Sons

From Theory to Practice in Soil Mechanics

Laurits Bjerrum

A. W. Skempton

The Terzaghi-school by no means allowed anyone in the field of soil mechanics to dispute von Terzaghi's right to stand as the founder of soil mechanics. In the beginning of 1962, an article appeared in the *Österreichische Ingenieur-Zeitschrift* in which a statement was quoted to the effect that O.K. Fröhlich was, together with von Terzaghi, the founder of soil mechanics. A. Casagrande reacted immediately. He persuaded Dr. Golder from Toronto, Canada, who was known as a brilliant writer with a special sense of humor, to write a letter to the leading journal in soil mechanics, *Geotechnique*, because he was able to "give to such a letter just the right touch of sting." Dr. Golder wrote such a letter to the Editor of Geotechnique, with the title "Claims of Fatherland," not mentioning, however, the name Fröhlich. In the last two passages he stated:

"For the sake of future historians of the subject, and as there will assuredly arise many kings who knew not David, cannot we who know him agree that there is no ambiguity of paternity in this case, although the identity of the mother remains in doubt unless this be fecund mother earth.

To continue my metaphor, the *conception* of soil mechanics was, in my opinion, the paper *Die Berechnung der Durchlässigkeitsziffer des Tones aus dem Verlauf der hydrodynamischen Spannungserscheinungen* in 1923, the *birth* was *Erdbaumechanik auf bodenphysikalischer Grundlage* in 1925 and the *coming of age* was 'From Theory to Practice in Soil Mechanics' in 1960. One man alone was responsible for this achievement, and no one could be happy who claimed to share it with him."

If one reads von Terzaghi's statements in the foregoing section carefully, it is not surprising that, for the creation of a theory of saturated and non-saturated porous solids, scientists in the field of soil mechanics have contributed little – with only few exceptions – as a result of the fact that von Terzaghi had an eminent influence on the development of soil mechanics. In particular in the USA, von Terzaghi was, and still is in certain circles, a legend. Von Terzaghi's view is in some ways hard to understand because it was completely clear to him that a saturated porous solid was in a certain sense a mixture; this view coincides with modern standards (von Terzaghi, 1925b). However, von Terzaghi did nothing to transform his ideas into theory. Moreover, his disciples – in particular, Arthur Casagrande – adhered to von Terzaghi's viewpoint. Thus, one can understand that, since then, many branches of soil mechanics have drifted into a direction of a purely experience-science, with no scientific foundation of its results. Of course, it should be emphasized that soil, as a natural material, is very complex, and thus in general is not open for an exact mechanical-mathematical treatment. However, that is not the point. Rather, a theory always deals with ideal materials and serves to recognize the assumptions and limits of any calculation of practical problems. This criticism is directed against the attempt on the part of representatives of soil mechanics to hinder the scientific investigation on the basis of well-founded mechanical axioms and mathematical rules.

6. Epilogue

After the death of von Terzaghi, his disciples did not hesitate to turn him into a legend. On the occasion of the 150th-anniversary of the Technische Hochschule of Vienna in 1965, A. Casagrande donated a bust of von Terzaghi to the Technische Hochschule. This bust was not unveiled, however, before the renaming of a lecture-hall as the *Terzaghi Lecture-Hall* on March 31, 1967. This was a big event at the Technische Hochschule and, in the evening, the Rector gave a reception for von Terzaghi's closest relatives living in Europe at that time.

A. Casagrande also made great efforts to establish the *Terzaghi Lecture.* A street in Vienna was even rechristened *Terzaghigasse.*

Ella Byloff (sitting) and Vera, 1967

Karl von Terzaghi Lecture Room at the Technische University of Vienna

"Terzaghi-Street" in the 22nd district of Vienna

In 1969, a memorial tablet was unveiled at von Terzaghi's birthplace, and further, at the house in Graz, a plaque reminds all to this day of the founder of modern soil mechanics, and the Mission Dam in British Columbia was renamed *Terzaghi Dam.*

In 1975, there was a conference in Vienna in honor of Karl von Terzaghi. At this conference, the efforts to praise von Terzaghi culminated in this speech by his wife:

"I believe that Karl's personality was accurately characterized by our good old friend of the thirties Professor Fritz Haas who said that Karl was a renaissance man. It was in the renaissance period that men first saw earth and sky in all their fresh and unexplored splendor, unclouded by tradition, myths or superstition. It was during those years of change and development that they first dared to give more credence to the witness of their own eyes than to the belief of past generations. Instead of consulting traditional authority they began to ask questions of nature. Although this liberation spread only slowly throughout the fabric of intellectual life and indeed is not yet complete, its impact was revolutionary, wherever it took place and men responded with excitement and with joy, as the domination of the unseen world began to fade and the appreciation of the pleasures of the real world increased. For the first time in perhaps a thousand years, at least a select few could expect to enjoy life for its own sake. In retrospect it appears that renaissance man stood on a divide, liberated from the scholasticism of the past but not yet dominated by the machines to which later generations would become addicted. Perhaps it was this freedom both from the burdens of the past and from those of the future that gave such an exceptionally large number of individuals the leisure as well as the creative urge to develop a broad spectrum of talents and interests. This was I believe an ambience that Karl would have found very congenial. He too occupied that divide to which man had ascended from the foggy valleys dominated by myth and from which they had not yet descended to a slavish addiction to gadgets. Karl had neither time nor interest for such distractions as radio, TV or automobiles. But throughout much of his life he found the leisure for capturing his delightful impressions of a landscape or a seascape with pencil and water color and he never failed to respond with appreciation and enthusiasm to the attainments of others in a wide variety of fields including literature, painting, music and last but by no means least the culinary and the wine making arts. He was born to doubt the validity of tradition and authority and he began by rejecting both the military school and the military career to which tradition assigned an officer's son. He felt this was a condemnation rather than a good tradition. His interest in Darwin's theory of evolution, which was then still anathema to many of his elders nearly led to his expulsion from high-school. Still later he was classified as a dangerous rebel who actually wished to substitute fact for fancy when he firmly rejected most of the rules of thumb which passed for engineering science throughout much of his career. In all these activities he shared with his renaissance forbears a zest for the world around him and a boundless joy in discovery."

Furthermore, the English engineer R. Glossop remarked in a review of *From Theory to Practice in Soil Mechanics* in the Journal *Géotechnique* **13** (1963) that "in the addition to those figures in the engineering world whose names are connected with some remarkable structure, there is a small group of great engineers, often scarcely known to the layman, who have profoundly influenced the mode of thought of their contempories, and consequently have initiated new eras in civil engineering practice. Professor Terzaghi is one such man, and his stature is comparable to that of Navier in the early 19th century, and Euler in the 18th century." Reading this quotation one can only state that R. Glossop obviously neither read the work of the great scientists Euler and Navier nor did he read von Terzaghi's work.

Finally, some remarks will be made concerning the fates of some people closely related to Karl von Terzaghi:

Karl's sister Ella passed away in 1978.

Only a few facts are known about Olga Byloff's stay in Mexico in 1912 and later. Karl von Terzagi neither mentioned explicitly the fate of the child Olga was pregnant with nor did he report on his daily life after his return to Graz in December 1913 untill the end of 1914.

Olga Ruchty (von Terzaghi, maiden name Byloff) moved in the middle of the 1950s to Garsten, Austria, to her daughter Vera and her son-in-law. She was taken ill with Parkinson's disease. At the very beginning of the 1960s she died.

Ruth von Terzaghi (Doggett), in the late 1960s involved in the protest against the Vietnam war, passed away in the early 1990s.

Vera lived with her mother in Graz untill 1934. Later she was trained as a nurse for children and after Austria's annexion she served as the leader of the BDM (Union of German Girls) of Styria. In 1952 she married the biologist Dr. Rößner, who was also associated with the Nazis. They lived mainly in Garsten, Austria. In June, 1981, Vera died of a cerebral apoplexy.

Karl's children with Ruth, Eric and Peggy, live today in New Zealand and Maine, USA.

His disciple Arthur Casagrande tried to make Karl von Terzaghi a legend till September 6, 1981, the day he died. And Lauritz Bjerrum, the highly gifted director of the Norwegian Geotechnical Institute in Oslo, suffered a deadly heart attack on February 27, 1973.

O. K. Fröhlich became Karl von Terzaghi's sucessor at the Technische Hochschule in 1940. He passed away on January 20, 1964.

Acknowledgements

Numerous people have given me the support to enable me to write this book. I would like to thank Dr. techn. A. Lechner, Dipl.-Ing. E. J. Jiresch and Mrs. Dr. J. Mikoletzky from the Archive of the Technical University of Vienna for their cooperation. In particular, I thank Dr. techn. A. Lechner for his assistance in finding the important inquiry report of the inquiry committee in 1988, which was still under seal at that time. I thank also the archivists in Graz from the University, Prof. Dr. A. Kernbauer, and the Technical University, Mrs. Dr. M. Vesulak, as well as from the University of Vienna, Mrs. B. Kromp, and the Staatsarchiv in Vienna, for their cooperation.

Mrs. M. Martin-Heaton of the Massachusetts Institute of Technology (M.I.T.) made documents available related to Karl von Terzaghi's stay at M.I.T. from 1925 to 1929.

The current custodian at the Terzaghi Library in Oslo, Mrs. W. Enersen has assisted me in every respect in finding the files, correspondence etc., important for Karl von Terzaghi's biography, and I would like to express my deep gratitude to her.

I was extremely fortunate to be able to correspond with Dr. Ruth Terzaghi, Karl von Terzaghi's second wife, and to interview her in her apartment in Winchester, Massachusetts, in 1990. D. Casagrande in Arlington, Massachusetts, placed all the correspondence of his uncle A. Casagrande with Karl von Terzaghi at my disposal and he and his mother helped with personal memories to reveal some unclear points in von Terzaghi's character.

Further, I have carried on correspondence and helpful interviews with von Terzaghi's relatives in Graz, Mrs. Elisabeth Byloff (daughter of Ella and Prof. F. Byloff), Mrs. G. Pongratz (von Terzaghi's granddaughter), Dr. H. Puchwein (son of Elisabeth Byloff and Prof. Puchwein), and Dr. H. Rößner (husband of Vera von Terzaghi). I would like to thank them all, in particular, Dr. H. Puchwein for providing some photographs and details in Karl von Terzaghi's life and Mrs. G. Pongraz for leaving the correspondence of Karl von Terzaghi with his daughter Vera to the author.

Several colleagues have given me valuable hints concerning the development of the theory of porous media and the biography of Karl von Terzaghi as well as the strange affair in Vienna in the 1930s. In particular, I would like to thank Professor Dr.-Ing. W. Ehlers, Professor Dr. Z. Liu and Professor Dr.-Ing. J. Bluhm for valuable and stimulating discussions over many years in Essen. Moreover, I profited from discussions and correspondence with the Professors R. Jelineck (former assistant of Karl von Terzaghi), E. F. Richart, E. Reißner, R. E. Goodman, K. Desoyer, and D. Flamm.

I feel deeply obliged to recall the late Professor R. Schiffman who was enthusiastically interested in the historical development of soil mechanics and, in particular, in the biography of Karl von Terzaghi; and it was a honor for me to write a paper together with him on the affair in Vienna.

The research work on the biography of Karl von Terzaghi and the scandal in Vienna could only have been carried out with the financial support of the Deutsche Forschungsgemeinschaft (DFG), the Volkswagen-Stiftung, the Forschungspool of the University of Essen, and the Fördervereinigung für die Stadt Essen. This support is gratefully acknowledged.

References

1794/99 Woltman, R.: Beyträge zur Hydraulischen Architectur, Dritter Band, Vierter Band, Johann C. Dietrich, Göttingen.

1871 Mohr, O.: Beitrag zur Theorie des Erddrucks, *Zeitschrift des Architekten- und Ingenieurvereines für das Königreich Hannover*, **17**, 344-372.

1871 Stefan, J.: Über das Gleichgewicht und die Bewegung, insbesondere die Diffusion von Gasmengen, *Kaiserliche Akademie der Wissenschaften in Wien, mathematisch-naturwissenschaftliche Klasse, Abteilung IIa* **63**, Wien, 63–124.

1872 Mohr, O.: Beitrag zur Theorie des Erddrucks, *Zeitschrift des Architekten- und Ingenieurvereines für das Königreich Hannover*, **18**, 246-248.

1885 Reynolds, O.: On the dilatancy of media composed of rigid particles in contect, with experimental illustrations, *Philosophical Magazine*, **20**, 469-481.

1893 Kötter, F.: Die Entwicklung der Lehre vom Erddruck, Jahresbericht der deutshen Mathematiker-Vereinigung, Berlin.

1906 Geikie, A.: Anleitung zu Geologischen Aufnahmen, German by Karl von Terzaghi, Franz Deuticke, Leipzig und Wien.

1912 Fillunger, P.: Drei wichtige ebene Spannungszustände, *Zeitschrift für Mathematik und Physik* **60**, 275-285.

Rudeloff; Panzerbieter: Versuche über den Porendruck des Wassers im Mauerwerk, Mitt. a. d. kgl. Materialprüfungsamt zu Gr. Lichterfeld, Erg. H. 1.

1913 Fillunger, P.: Der Auftrieb in Talsperren, *Österreichische Wochenschrift für den öffentlichen Baudienst* (Ö W Ö B) **19**, 532–556, 567–570.

1914 Fillunger, P.: Neuere Grundlagen für die statische Berechnung von Talsperren, *Zeitschrift des Österreichischen Ingenieur- und Architekten-Vereines* (Z ö I A V), **23**, 441–447.

1915 Fillunger, P.: Versuche über die Zugfestigkeit bei allseitigem Wasserdruck, *Österreichische Wochenschrift für den öffentlichen Baudienst* (Ö W Ö B) **29**, 443–448.

1918a Fillunger, P.: Die Berechnung genieteter Vollwanträger, *Sitzungsberichte der Akademie der Wissenschaften in Wien, mathematisch-naturwissenschaftliche Klasse, Abteilung IIa* **127**, 1987–2051.

b Fillunger, P.: Zur Festigkeitsberechnung von Furniertragflächen, *Österreichische Flugzeitschrift*, Heft 13/14.

c Fillunger, P.: Theorie der Spornabfederung von Flugzeugen, *Österreichische Flugzeitschrift*, Heft 23/24.

1919 von Terzaghi, K.: Die Erddruckerscheinungen in örtlich beanspruchten Schüttungen und die Entstehung von Tragkörpern, *Österreichische Wochenschrift für den öffentlichen Baudienst*, Nos. 17–19, pp. 194–199, 206–210, 218–223.

1920 von Terzaghi, K.: Old earth-pressure theories and new test results, *Engineering News-Record*, **25**, No. 14, pp.632–634.

1922 von Terzaghi, K.: Der Grundbruch an Stauwerken und seine Verhütung, *Die Wasserkraft, Zeitschrift für die gesamte Wasserwirtschaft*, **17**, 445–449.

1923 von Terzaghi, K.: Die Berechnung der Durchlässigkeitsziffer des Tones aus dem Verlauf der hydrodynamischen Spannungerscheinungen. *Sitzungsberichte der Akademie der Wissenschaften in Wien, mathematisch-naturwissenschaftliche Klasse, Abteilung IIa* **132**, (No 3/4), 125–138.

1924 Kozeny, J.: Über den kapillaren Aufstieg des Wassers im Boden, *Kulturtechnik*, Heft 1, 27, 11-16.

1925a von Terzaghi, K.: Erdbaumechanik auf bodenphysikalischer Grundlage, Franz Deuticke, Leipzig/Wien.

b von Terzaghi, K.: Principles of soil mechanics, *Engineering News-Record* **19**, 742–746, **20**, 796–800, **21**, 832–936, **22**, 974–978, **23**, 912–915, **25**, 987–999, **26**, 1026–1029, **27**, 1064–1068.

1927 Kozeny, J.: Über kapillare Leitung des Wassers im Boden (Aufstieg, Versickerung und Anwendung auf die Bewässerung), *Sitzungsberichte der Akademie der Wissenschaften in Wien, mathematisch-naturwissenschaftliche Klasse, Abteilung IIa* **136**, 271–309.

Krey, H.: Rutschgefährliche und fließende Bodenarten, *Die Bautechnik* **5**, 485–489.

1928a Fillunger, P.: Die Leitgedanken der statischen Berechnung von Flugzeugen, *Militärwissenschaftliche technische Mitteilungen*, Heft Juli–Oktober.

b Fillunger, P.: Zur Frage der Betonpfahlgründungen, Zeitschrift des Östereichischen Ingenieur- und Architekten-Vereines (Z ö I A V), Heft 41/42, 1927/28.

Stern, O.: Zur Frage der Betonpfahlgründungen, *Zeitschrift des Österreichischen Ingenieur- und Architekten-Vereines* (Z ö I A V), Heft 43/44, 416, Heft 47/48, 452.

1929 Hoffman, O.: Zur Frage des Auftriebs in Talsperren, *Wasserwirtschaft* **22**, 562–566.

1930 Ortenblad, A.: Mathematical theory of the process of consolidation of mud deposits, *Journal of Mathematics and Physics* **9**, no. 2, 73–149.

1933 von Terzaghi, K.: Auftrieb und Kapillardruck an betonierten Talsperren, *Wasserwirtsch.* **26**, 397–399.

1934 Lehr, E.: Spannungsverteilung in Konstruktionselementen, V.D.I.-Verlag, Berlin.

Fillunger, P.: Der Kapillardruck in Talsperren. *Wasserwirtschaft* **27**, Heft 13/14.

von Terzaghi, K.: Large Retaining-Wall Tests, I–V, *Engineering News Record*, **112**, 136-140, 259-262, 316-318, 403-406, 503-508, with corrections on p. 747.

von Terzaghi, K. and Rendulic, L.: Die wirksame Flächenporosität des Betons, *Zeitschrift des Österreichischen Ingenieur- und Architekten Vereines* (Z Ö I A V), Heft 1/2 , 1-9.

1935 Biot, M. A.: Le problème de la consolidation des matières argileuses sous une charge, *Annales de la Societé scientifique de Bruxelles* (Ann Soc sci Brux) **B 55**, 110–113.

Fillunger, P.: Alte und neue Probleme der Festigkeitslehre, Selbstverlag des Verfassers, Wien.

Rendulic, L.: Der hydrodynamische Spannungsausgleich in zentral entwässerten Tonzylindern, *Wasserwirtschaft und Technik*, Heft 23/24, 250–253, 269–273.

1936 Fillunger, P.: Erdbaumechanik?, Selbstverlag des Verfassers, Wien.

von Terzaghi, K.: Fröhlich, O. K.: Theorie der Setzung von Tonschichten, Franz Deuticke, Leipzig /Wien.

Rendulic, L.: Porenziffer und Porenwasserdruck in Tonen, *Der Bauingenieur* **27**, 559–564.

Casagrande, A. (ed.): Proceedings of the First International Conference on Soil Mechanics and Foundation Engineering, Harward University, Cambridge, Mass., Vol. I, II, III.

1937 Fillunger, P.: Erdbaumechanik und Wissenschaft: Eine Erwiderung (ed. Erwin Fillunger), Selbstverlag, Wien.

von Terzaghi, K.: Fröhlich, O. K.: Erdbaumechanik und Baupraxis: Eine Klarstellung, Franz Deuticke, Leipzig • Wien.

1938 Flamm, L.: Beitrag zur Theorie der Setzung von Tonschichten, *Wasserkraft und Wasserwirtschaft* (Wass Kr) **33**, 97–98.

Heinrich, G.: Wissenschaftliche Grundlagen der Theorie der Setzung von Tonschichten, *Wasserkraft und Wasserwirtschaft* (Wass Kr) **33**, 5–10.

Kögler, F. and Scheidig, A.: Baugrund und Bauwerk, Verlag von Wilhelm Ernst & Sohn, Berlin.

1942 Carrillo, N.: Simple two and three dimensional cases in the theory of consolidation of soils, *Journal of Mathematics and Physics* **21**, 1–5.

1943 von Terzaghi, K.: Theoretical Soil Mechanics, John Wiley & Sons, New York.

1948 von Terzaghi, K. and Peck, R. B.: Soil Mechanics in Engineering Practice, John Wiley & Sons, New York.

1955 Heinrich, G. and Desoyer, K.: Hydromechanische Grundlagen für die Behandlung von stationären und instationären Grundwasserströmungen, *Ingenieur-Archiv* (Ing-Arch) **23**, 73–84.

1956 Heinrich, G. and Desoyer, K.: Hydromechanische Grundlagen für die Behandlung von stationären und instationären Grundwasserströmungen, II. Mitteilung, *Ingenieur-Archiv* (Ing-Arch) **24**, 81–84.

1958 Mann, G.: Deutsche Geschichte des 19. und 20. Jahrhunderts, S. Fischer, Berlin.

1960 Skempton, A. W.: Significance of Terzaghi's concept of effective stress (Terzaghi's discovery of effective stress) in: From theory to practice in soil mechanics (eds. Bjerrum, L., Casagrande, A., Peck, R. B., Skempton, A. W.), Wiley, New York • London.

1961 Heinrich, G. and Desoyer, K.: Theorie dreidimensionaler Setzungsvorgänge in Tonschichten, *Ingenieur-Archiv* (Ing-Arch) **30**, 225–253.

1973 Glossop, R.: 'The influence of Terzaghi on civil engineering practice in England, in: Terzaghi Memorial Lectures (eds.: S.S Rerzcan, A. S. Yalçin), Bogāzoçi University Publications, Istambul.

1979 Krause, P.: "O alte Burschenherrlichkeit", Die Studenten und ihr Brauchtum, Verlag Styria, Graz • Wien • Köln.

1994 Popper, K. R.: Alles Leben ist Problem lösen – Über Erkenntnis, Geschichte und Politik, Piper, München und Zürich.

1999 Goodman, R. E.: Karl Terzaghi – The Engineer as Artist, Copyright by the American Society of Civil Engineers (ASCE Press).

2000 de Boer, R.: Theory of Porous Media – Highlights in the Historical Development and Current State, Springer, Berlin • Heidelberg • New York.

Selected documents (copies)

which has been used in this treatise, however, has not been cited in many cases:

Karl von Terzaghi's diaries from 1900 – 1939 (Terzaghi Library, Oslo; University of Essen)[64].
Karl von Terzaghi's letters to his grandfather, his mother and Prof. Wittenbauer (Terzaghi Library, Oslo).
Correspondence of Karl von Terzaghi with Arthur Casagrande (partly at the University of Essen).
Karl von Terzaghi's letters to his daughter Vera (University of Essen).
Inquiry report (Gutachten mit Beilagen) (Archive of the University of Vienna).
Karl von Terzaghi: Mein Lebensweg und meine Ziele (1932) (Archive of the Technical University of Vienna).
Karl von Terzaghi: Bericht an den österreichischen Ingenieur- und Architektenverein, Wien (March 31,1912) (University of Essen).
Karl von Terzaghi: Near East Recollections (Terzaghi Library, Oslo).
Correspondence of O.K. Fröhlich (Archive of the University of Vienna).
Various newspapers from around March 8, 1937 (Archive of the University of Vienna).
Karl von Terzaghi: Sketches from return to Germany from USA and a trip to Soviet Russia 1929 (Terzaghi Library, Oslo).
Personal files of Karl von Terzaghi and Paul Fillunger (Archive of the Technical University of Vienna).
Karl von Terzaghi: Ziele und Wege des Sowjetsystems, *Wochenschrift des Niederösterreichischen Gewerbevereins* (April 30, 1931).
Karl von Terzaghi's correspondence with various persons (Archive of the Technical University of Vienna, Terzaghi Library, Oslo).
Letters of Karl von Terzaghi to Paul Fillunger (Archive of the Technical University of Vienna).
Karl von Terzaghi: Kontroverse mit Professor Fillunger und ihr tragischer Ausgang, Winter 1936/37 (Terzaghi Library, Oslo).
Karl von Terzaghi: Lebenslauf 1932–1938 (Terzaghi Library, Oslo).
Karl von Terzaghi: Memorandum betreffend Vorgeschichte und sachlicher Hintergrund des Angriffes – mit vier Beilagen, A bis D (Terzaghi Library, Oslo).
Einladung und Protokoll der Disziplinarkammer für Bundeslehrer an der Technischen Hochschule in Wien (Archive of the Technical University of Vienna).

[64] Notes in the parentheses are advices, where the documents can be found.

Einvernahmen (questionings) of Fillunger, von Terzaghi and other Viennese professors by Dr. Goldberg (Archive of the Technical University of Vienna).

Selected photos

Copies of photos were taken from the report: Brandl, H.: Mitteilungen für Grundbau, Bodenmechanik und Felsbau (Technische Universität Wien), Heft 2, Wien (1983–1984) and from the book by Goodman (1999) (see References) as well as from: Bodo Harenberg: Chronik 1914, Chronik Verlag im Bertelsmann Lexikon Verlag and from: Schütz, E. and Gruber, E.: Mythos Reichsautobahn, Ch. Links Verlag, Berlin (1996).

Most of the photos stem from the Terzaghi Library, Oslo, the Archive of the Technical University of Vienna and the Archive of the University of Vienna.

Personal photos were provided to the author by the late Ruth von Terzaghi, by relatives of Karl von Terzaghi in Austria, in particular by Dr. Puchwein, and by Harald Fillunger.

Author Index

Subject Index

MIX
Papier aus verantwortungsvollen Quellen
Paper from responsible sources
FSC® C105338

If you have any concerns about our products,
you can contact us on
ProductSafety@springernature.com

In case Publisher is established outside the EU,
the EU authorized representative is:
Springer Nature Customer Service Center GmbH
Europaplatz 3, 69115 Heidelberg, Germany

Printed by Libri Plureos GmbH
in Hamburg, Germany